Distribution Solutions of Nonlinear Systems of Conservation Laws

MEMOIRS
of the
American Mathematical Society

Number 889

Distribution Solutions of Nonlinear Systems of Conservation Laws

Michael Sever

November 2007 • Volume 190 • Number 889 (second of three numbers) • ISSN 0065-9266

American Mathematical Society
Providence, Rhode Island

2000 *Mathematics Subject Classification.*
Primary 35L65, 35L67.

Library of Congress Cataloging-in-Publication Data

Sever, Michael, 1941–
Distribution solutions of nonlinear systems of conservation laws / Michael Sever.
p. cm. — (Memoirs of the American Mathematical Society, ISSN 0065-9266 ; no. 889)
"Volume 190, number 889 (second of three numbers)."
Includes bibliographical references.
ISBN 978-0-8218-3990-4 (alk. paper)
1. Conservation laws (Mathematics) 2. Nonlinear theories. I. Title.

QA377.S4642 2007
515′.3535—dc22 2007060778

Memoirs of the American Mathematical Society

This journal is devoted entirely to research in pure and applied mathematics.

Subscription information. The 2007 subscription begins with volume 185 and consists of six mailings, each containing one or more numbers. Subscription prices for 2007 are US$649 list, US$519 institutional member. A late charge of 10% of the subscription price will be imposed on orders received from nonmembers after January 1 of the subscription year. Subscribers outside the United States and India must pay a postage surcharge of US$38; subscribers in India must pay a postage surcharge of US$43. Expedited delivery to destinations in North America US$53; elsewhere US$130. Each number may be ordered separately; *please specify number* when ordering an individual number. For prices and titles of recently released numbers, see the New Publications sections of the *Notices of the American Mathematical Society.*

Back number information. For back issues see the *AMS Catalog of Publications.*

Subscriptions and orders should be addressed to the American Mathematical Society, P. O. Box 845904, Boston, MA 02284-5904, USA. *All orders must be accompanied by payment.* Other correspondence should be addressed to 201 Charles Street, Providence, RI 02904-2294, USA.

Memoirs of the American Mathematical Society is published bimonthly (each volume consisting usually of more than one number) by the American Mathematical Society at 201 Charles Street, Providence, RI 02904-2294, USA. Periodicals postage paid at Providence, RI. Postmaster: Send address changes to Memoirs, American Mathematical Society, 201 Charles Street, Providence, RI 02904-2294, USA.

This publication is indexed in *Science Citation Index*®, *SciSearch*®, *Research Alert*®, *CompuMath Citation Index*®, *Current Contents*®*/Physical, Chemical & Earth Sciences.*
Printed in the United States of America.

♾ The paper used in this book is acid-free and falls within the guidelines established to ensure permanence and durability.
Visit the AMS home page at `http://www.ams.org/`

10 9 8 7 6 5 4 3 2 1 12 11 10 09 08 07

The author expresses the warmest thanks to Professors Richard Saunders and Barbara Keyfitz, without whose help this work would not have been possible.

Contents

ABSTRACT. The local structure of solutions of initial value problems for nonlinear systems of conservation laws is considered. Given large initial data, there exist systems with reasonable structural properties for which standard entropy weak solutions cannot be continued after finite time, but for which weaker solutions, valued as measures at a given time, exist. At any given time, the singularities thus arising admit representation as weak limits of suitable approximate solutions in the space of measures with respect to the space variable.

Two distinct classes of singularities have emerged in this context, known as delta-shocks and singular shocks. Notwithstanding the similar form of the singularities, the analysis of delta-shocks is very different from that of singular shocks, as are the systems for which they occur. Roughly speaking, the difference is that for delta-shocks, the density approximations majorize the flux approximations, whereas for singular shocks, the flux approximations blow up faster. As against that admissible singular shocks have viscous structure.

2000 *Mathematics Subject Classification.* 35L65, 35L67

Key words and phrases. systems of conservation, delta-shocks, singular shocks, distribution solutions

Received by the editor February 10, 2003.

CHAPTER 1

General distribution solutions

1. Something worse can happen

We are concerned here with the solution of initial value problems for nonlinear, first order systems of conservation laws, of the form

$$(1.1) \qquad w_t + q(w)_x = 0, \quad x \in \mathbb{R}, t > 0$$

$$(1.2) \qquad w(x,t) \in D \subseteq \mathbb{R}^n \quad \text{a.e. in } \mathbb{R} \times \mathbb{R}_+$$

$$(1.3) \qquad q \in C^2 : D \to \mathbb{R}^n$$

with D a given open set, and initial data

$$(1.4) \qquad w(\cdot, 0) : \mathbb{R} \to D.$$

For $w(\cdot,0)$ smooth and bounded, and bounded uniformly away from ∂D, the systems we consider admit unique smooth solutions, continuously dependent on the initial data, for some finite time depending on the initial data. In particular the characteristic speeds, the eigenvalues of $q_w(w)$, are real for all $w \in D$. It is well-known that in general for q nonlinear, the continuation of a solution depends on the acceptance of weak solutions, involving a loss of regularity in w and a related fundamental uniqueness question.

A natural question is whether anything worse can happen. In his signal paper [G], James Glimm found weak solutions of (1.1) as discontinuous functions of bounded variation in x, pointwise in t, satisfying (1.1) in the space of measures on $\mathbb{R} \times \mathbb{R}_+$, and continuing indefinitely in t. He suggested: "Presumably nothing worse can happen. Apparently the concept of weaker solutions, e.g. distributions, has no meaning because q is not linear."

When initial data of large variation is permitted, however, the lack of a maximum principle for systems (1.1) indicates that worse things can indeed happen. Consider a system (1.1) with initial data satisfying

$$w_1(x,0) \geq \underline{w_1} > 0, x \in \mathbb{R}$$

and admitting a unique solution, either a classical solution or an entropy weak solution, up to time $t_1 > 0$ but such that for some $t \in (0, t_1]$ $w_1(\cdot, t)$ assumes negative values on a set of finite measure in $\mathbb{R}$. This is very mild requirement, taking linear combinations of the equations in (1.1) and adding constants to w as necessary. We now regard x as a "Lagrangian" space coordinate, and introduce an "Eulerian" space coordinate Y determined from

$$(1.5) \qquad dY = w_1 dx - q_1 dt.$$

The corresponding "Eulerian" system is given by

$$\begin{aligned}&\left(\frac{1}{w_1}\right)_t - \left(\frac{q_1}{w_1}\right)_Y = 0\\ &\left(\frac{w_j}{w_1}\right)_t + \left(q_j - \frac{q_1}{w_1}w_j\right)_Y = 0, \quad j = 2, \dots, n.\end{aligned} \tag{1.6}$$

For $w_1 > 0$, the system (1.6) inherits whatever structural assumptions (strict hyperbolicity, genuine nonlinearity, existence of a strictly convex entropy, form of the Hugoniot locus, etc.) [Ge] as those for the system (1.1); indeed, weak solutions of (1.1) and (1.6) are isomorphic [W1, W2]. But (1.6), with the corresponding initial data, is clearly not solvable in the class of bounded measurable entropy weak solutions up to time t_1, as this would imply a second such solution of (1.1), with w_1 remaining positive. This is reconciled with Glimm's result by observing that the corresponding initial data for the system (1.6) is necessarily not of small variation.

Additional insight into the problem is obtained by observing that the same argument applies to the class of self-similar entropy weak solutions of the systems (1.1) as (1.6), in the variables x/t or Y/t respectively. So Riemann data for (1.1) with w_1 positive but such that negative value(s) of w_1 are obtained in the intermediate states of the corresponding self-similar entropy solution correspond to unsolvable Riemann problems in this class for the system (1.6) with the corresponding data. Again this is resolved with the classical results on solvability of Riemann problems [L1] by noting that the Riemann data for the unsolvable problem (1.6) is not of small variation.

Fairly strong results exist [DD, Liu1, Liu2, Liu3, S1, KK7] on the solvability of large data Riemann problems for (1.1) (or (1.6)) assuming that the Hugoniot locus of any given point $w_0 \in D$, given by

$$\begin{aligned}\Gamma(w_0) =& \{w \in D/\{w_0\} | q(w) - q(w_0) = s(w, w_0)(w - w_0)\}\\ & s(w, w_0) \in \mathbb{R}\end{aligned} \tag{1.7}$$

contains $2n$ or distinct 1-manifolds connecting w_0 with infinity or possibly with ∂D. Thus we may broadly (if not exclusively) identify unsolvable Riemann problems with a Hugoniot locus which is deficient in this respect. Indeed, all of the published examples of distribution solutions of systems (1.1) correspond to systems with classically unsolvable Riemann problems, and most correspond to systems in which either $\Gamma(w_0)$ has too few disjoint branches in a neighborhood of w_0 or else the branches do not continue indefinitely.

In this context, inability to continue an entropy weak solution of (1.1) can obviously be caused by a collision of two or more entropy shocks such that the resulting Riemann problem is unsolvable. It can also happen that smooth solutions simply blow up (or reach ∂D). This latter phenomenon is usually associated with nonhyperbolic systems such as "zero pressure gas dynamics"; however, here is an example of such behavior where the system is strictly hyperbolic.

Consider a pair of conservation laws (1.1) which is strictly hyperbolic, $\lambda_-(w) < \lambda_+(w)$ for all $w \in D$, where

$$q_w(w)r_\pm(w) = \lambda_\pm(w)r_\pm(w), \quad r_\pm(w) \in \mathbb{R}^2/\{0\} \tag{1.8}$$

such that $\lambda_+(w)$, $\lambda_-(w)$ uniquely determine w, and such that both characteristic families are linearly degenerate,

$$r_\pm(w) \cdot \lambda_{\pm,w}(w) = 0. \tag{1.9}$$

From (1.8), (1.9), $\lambda_\pm$ are also Riemann invariants, satisfying

$$\lambda_{\pm,w}(w)q_w(w) = \lambda_\mp(w)\lambda_{\pm,w}(w),$$

and thus for smooth solutions, using (1.1),

$$\lambda_\pm(w)_t + \lambda_\mp(w)\lambda_\pm(w)_x = 0. \tag{1.10}$$

From (1.9), these systems are of Temple class [T1, LT, Se1, Se2], explicitly solvable as long as a solution exists. However the pair (1.10) is discussed in [SV], where it is shown for smooth data that in general $\lambda_+ - \lambda_-$ decreases to zero in finite time. In the present context that implies that w blows up in finite time. A simplified proof of these results for this specific example follows.

The pair (1.10) is not in conservation form; introducing

$$u = \frac{1}{2}(\lambda_+ + \lambda_-),\ v = \frac{1}{2}(\lambda_+ - \lambda_-) \tag{1.11}$$

and using (1.10), we find

$$\begin{aligned} u_t + uu_x - vv_x &= 0 \\ v_t + uv_x - vu_x &= 0. \end{aligned} \tag{1.12}$$

Regarding w, q as functions of u, v in (1.1), using (1.12) we find for smooth solutions of (1.1)

$$(-uw_u + vw_v + q_u)u_x + (vw_u - uw_v + q_v)v_x = 0; \tag{1.13}$$

since u_x, v_x are arbitrary, this can hold only if the expressions in parentheses in (1.13) vanish separately,

$$q_u = uw_u - vw_v,\ q_v = -vw_u + uw_v. \tag{1.14}$$

Eliminating q from (1.14) we find that w satisfies the spherically symmetric wave equation in the u, v coordinates

$$w_{uu} = \frac{1}{v^2}(v^2 w_v)_v \tag{1.15}$$

for v positive.

Thus each component of w in (1.1), regarded as a function of u, v, is a solution of (1.15), with the corresponding component of q determined from (1.14). Any two such solutions of (1.15) may be chosen, provided that the two components of w uniquely determine u, v. Indeed, with both fields linearly degenerate, in the context of ordinary weak solutions it does not matter which solutions of (1.15) are adopted, as all conservation laws obtained in this manner will be satisfied weakly.

Therefore whatever system (1.1) was given is equivalent, for smooth and ordinary weak solutions, to the system

$$\begin{aligned} \left(\frac{1}{v}\right)_t + \left(\frac{u}{v}\right)_x &= 0 \\ \left(\frac{u}{v}\right)_t + \left(\frac{u^2}{v} - v\right)_x &= 0, \end{aligned} \tag{1.16}$$

obtained from the specific choices $w = 1/v$ and $w = u/v$ in (1.15).

Introducing a Lagrangian space coordinate y, satisfying a judiciously chosen special condition

$$dy = (dx - udt)/v \tag{1.17}$$

the system (1.16) is equivalent for positive v to the elementary wave equation,

$$\begin{aligned} v_t - u_y &= 0 \\ u_t - v_y &= 0 \end{aligned} \tag{1.18}$$

and it is clear that initially smooth and uniformly positive v can decrease to zero and negative values in finite time.

Once v decreases to zero at some point, the systems (1.16) and (1.18) are no longer equivalent; one questions whether any meaningful continuation of the solution of (1.16) is possible. Evidence that this is at least sometimes so obtains from consideration of weakly convergent sequences of approximate solutions, for which v remains positive.

As an example, consider Riemann initial data for the systems (1.16) or (1.18), of the form

$$\begin{pmatrix} u \\ v \end{pmatrix}(\cdot, 0) = \begin{cases} \begin{pmatrix} u_- \\ v_0 \end{pmatrix}, & x < 0(\text{or } y < 0) \\ \begin{pmatrix} u_+ \\ v_0 \end{pmatrix}, & x > 0(\text{or } y > 0),\ x, y \in \mathbb{R} \end{cases} \tag{1.19}$$

with $v_0 > 0, u_- > u_+$ and satisfying a judiciously chosen special condition

$$\frac{1}{4}(u_- - u_+)^2 - v_0(u_- - u_+) = v_0^2. \tag{1.20}$$

The corresponding solution of (1.18) is

$$\begin{pmatrix} u \\ v \end{pmatrix}(y, t) = \begin{cases} \begin{pmatrix} u_- \\ v_0 \end{pmatrix}, & y < -t \\ \begin{pmatrix} \frac{u_- + u_+}{2} \\ -\sqrt{2}v_0 \end{pmatrix}, & -t < y < t \\ \begin{pmatrix} u_+ \\ v_0 \end{pmatrix}, & y < t. \end{cases} \tag{1.21}$$

There is no classical solution of (1.16) corresponding to such data satisfying (1.20), but for any $\varepsilon > 0$ there is an approximate solution, self-similar in x/t, with v positive

$$\begin{pmatrix} u^\varepsilon \\ v^\varepsilon \end{pmatrix}(x, t) = \begin{cases} \begin{pmatrix} u_- \\ v_0 \end{pmatrix}, & x/t < \frac{u_- + u_+}{2} - \varepsilon \frac{u_- - u_+}{2v_0} \\ \begin{pmatrix} \frac{u_- + u_+}{2} \\ \varepsilon \end{pmatrix}, & \frac{u_- + u_+}{2} - \varepsilon \frac{u_- - u_+}{2v_0} < x/t < \frac{u_- + u_+}{2} + \varepsilon \frac{u_- - u_+}{2v_0} \\ \begin{pmatrix} u_+ \\ v_0 \end{pmatrix}, & \frac{u_- + u_+}{2} + \varepsilon \frac{u_- - u_+}{2v_0} < x/t. \end{cases} \tag{1.22}$$

Comparing (1.16) with (1.1), we identify

$$w^\varepsilon = \begin{pmatrix} 1/v^\varepsilon \\ u^\varepsilon/v^\varepsilon \end{pmatrix}, q^\varepsilon = \begin{pmatrix} u^\varepsilon/v^\varepsilon \\ (u^\varepsilon)^2/v^\varepsilon - v^\varepsilon \end{pmatrix} \tag{1.23}$$

pointwise in x, t; then for any $\theta \in C_0^1 = \mathbb{R} \times \mathbb{R}_+ \to \mathbb{R}^2$ an elementary computation, omitted here in the interest of brevity, shows that $w^\varepsilon, q^\varepsilon$ approximately satisfy (1.1)

$$\int\int \theta \cdot (w_t^\varepsilon + q_x^\varepsilon) dx dt = O(\varepsilon). \tag{1.24}$$

Furthermore, the approximations w^ε satisfy an entropy inequality [L2], which is perhaps surprising as both fields are linearly degenerate for the system (1.16). But this system (1.16) admits a strictly convex (in $1/v, u/v$) entropy density U with corresponding flux F,

$$U(w) = \frac{u^2}{2v} + \frac{v}{2}, F(w) = \frac{u^3}{2v} - \frac{uv}{2}, \tag{1.25}$$

and for $\theta \in C_0^1 : \mathbb{R} \times \mathbb{R}_+ \to \mathbb{R}_+$ using (1.22) and (1.20) we find (again omitting a straightforward computation)

$$\begin{aligned} &\int\int \theta(U(w^\varepsilon)_t + F(w^\varepsilon)_x dx dt \\ &\qquad = -\frac{1}{2}(u_- - u_+)^2 \int \theta(\frac{u_- + u_+}{2} t, t) dt + O(\varepsilon). \end{aligned} \tag{1.26}$$

Finally, as $\varepsilon \downarrow 0$ the functions $w^\varepsilon, q^\varepsilon$ converge weakly in the space of measures on $\mathbb{R}$, pointwise with respect to t, to respective limits w^0, q^0 given by

$$\begin{aligned} w^0 &= \begin{cases} \begin{pmatrix} 1/v_0 \\ u_-/v_0 \end{pmatrix}, x/t < \frac{u_- + u_+}{2} \\ \qquad + \begin{pmatrix} 1 \\ \frac{u_- + u_+}{2} \end{pmatrix} \frac{(u_- - u_+)}{v_0} t\delta(x - \frac{u_- + u_+}{2} t) \\ \begin{pmatrix} 1/v_0 \\ u_+/v_0 \end{pmatrix}, x/t > \frac{u_- + u_+}{2} \end{cases} \\ q^0 &= \begin{cases} \begin{pmatrix} u_+/v_0 \\ \frac{u_-^2}{v_0} - v_0 \end{pmatrix}, x/t < \frac{u_- + u_+}{2} \\ \qquad + \left(\frac{u_- + u_+}{2}\right) \begin{pmatrix} 1 \\ \frac{u_- + u_+}{2} \end{pmatrix} \frac{(u_- - u_+)}{v_0} t\delta(x - \frac{u_- + u_+}{2} t) \\ \begin{pmatrix} u_+/v_0 \\ \frac{u_+^2}{v_0} - v_0 \end{pmatrix}, \frac{x}{t} > \frac{u_- + u_+}{2} \end{cases} \end{aligned} \tag{1.27}$$

From (1.24), $w_t^0 + q_x^0 = 0$ in the dual space of C^1 functions of bounded support. However, as $q(w^0)$ is not defined, a significant extension of the concept of a weak solution of (1.1) is needed before w^0 can be accepted as a distribution solution of (1.16) for the data (1.19), (1.20).

In the following discussion we characterize systems for which such extension of the concept of a weak solution is possible and/or necessary, in the content of solving initial value problems. Two such extensions are described in detail. While the two extensions are incompatible with one another and differ in numerous elements of the respective theories, they lead to distribution solutions of (1.1) of the same form; the singularities which develop are known in the literature as "delta-shocks" and "singular shocks", respectively.

Several examples of each type are discussed in detail, intending to demonstrate that the appearance of distribution solutions is more widespread than the literature perhaps suggests. For example, the above example shows that delta-shocks can occur in strictly hyperbolic systems (1.1). In Chapter 3 we show how singular shocks can occur for systems which are not hyperbolic. An example in Chapter 2 extends the possible form of the function q for delta-shock systems.

2. Background

The general question to be addressed here is familiar enough: we have an initial value problem, in this case of the form (1.1-1.4), and an admissible solution w up to some positive time t_1, but which cannot be continued thereafter in the same class. Hope for continuation in an extended class of solutions arises from the existence of a convergent sequence of approximations $\{w^\varepsilon\}, \varepsilon > 0$, which continue to later times (often indefinitely) and achieve values in D.

The w^ε are approximate solutions, satisfying

$$w_t^\varepsilon + q(w^\varepsilon)_x \rightharpoonup 0 \quad \text{as } \varepsilon \downarrow 0 \tag{2.1}$$

in the sense of distributions. For a system (1.1) admitting a convex entropy density, the w^ε will be assumed to satisfy the corresponding entropy inequality.

For $t < t_1$, the $w^\varepsilon(\cdot, t)$ converge in such a manner that our solution w is recovered

$$w^\varepsilon(\cdot, t) \to w(\cdot, t), \tag{2.2}$$

$$q(w^\varepsilon(\cdot, t)) \to q(w(\cdot, t)), \ t < t_1. \tag{2.3}$$

For $t > t_1$ (or $t \geq t_1$) the $w^\varepsilon(\cdot, t)$ converge "weakly" thus determining a candidate $w^0(\cdot, t)$ to be considered as the continuation of a solution,

$$w^\varepsilon(\cdot, t) \rightharpoonup w^0(\cdot, t) \quad \text{as } \varepsilon \downarrow 0, \ t \geq t_1 \tag{2.4}$$

but such that for $t > t_1$ the weak limit $q^0(\cdot, t)$ is such that

$$q^0(\cdot, t) \stackrel{\text{def}}{=} \lim_{\varepsilon \downarrow 0} q(w^\varepsilon(\cdot, t)) \neq q(w^0(\cdot, t)). \tag{2.5}$$

In (2.5), the existence of the weak limit as $\varepsilon \downarrow 0$ follows from (2.1) and (2.4) and indeed with w^0 obtained from (2.4) for $t \geq t_1, (w^0(\cdot, t) \stackrel{\text{def}}{=} w(\cdot, t), t < t_1)$ and q^0 from (2.5), $(q^0(\cdot, t) \stackrel{\text{def}}{=} q(w(\cdot, t), t < t_1)$

$$w_t^0 + q_x^0 = 0 \tag{2.6}$$

in the sense of distributions. As against that, the right-hand term in (2.5) is likely not defined, as for $t > t_1, w(\cdot, t)$ is likely a distribution on $\mathbb{R}$, achieving values in D almost everywhere. It is in the sense of (2.5) that we refer to the "weak" convergence of the $\{w^\varepsilon\}$ in (2.4).

Therefore more precise questions to be addressed here include the identification of systems for which such a limit (2.4) exists; the form and regularity of the limit so obtained; the sense in which such a limit may be considered a solution of (1.1), in view of (2.5), (2.6); and the solvability of initial value problems of the form (1.1-1.4), given the additional solutions thus obtained.

Several specific features of nonlinear systems of conservation laws encourage such a discussion. First and foremost, there are examples of systems arising in reasonably honest applications requiring distribution-valued solution [AH, CCL, CL1, KK1, KPS, LC, Ro, SZ]. Such systems also arise as limits of Euler and Boltzmann systems [B, ERS, Li, MMZ].

Of course our discussion is based on generalization from the existing examples, for example with regard to the structural assumptions made for the system (1.1) and with regard to its symmetry group. Additionally, the entropy condition imposed on solutions or approximate solutions is typically motivated by the available examples, and the entropy condition imposed on the approximate solutions will have a major effect on the form of distribution solutions obtained.

Second, there are robust methods, such as discretization and regularization, for obtaining and describing the approximate solutions needed. The insight and guidance obtained from computations has, for example, been of major importance in the treatment of singular shocks [SS]. Third, systems of conservation laws often admit nontrivial symmetries, or else are perturbations of systems which do [BK, S6, S7, S8]. Such symmetries, and the corresponding invariant solutions, are essential in the discussion of singular shocks in Chapter 3.

Finally given a system (1.1), there are generally related systems, equivalent for some but not all solutions. Examples include the systems obtained by change of space coordinate and those obtained by exchange of conserved quantities [Ke, S2, S3].

Perhaps not surprisingly, the differences in the solution sets of such related systems are greater for distribution solutions than for ordinary weak solutions. The example of section 1 already indicates this phenomenon, first is the nonequivalence of the solutions of the corresponding Eulerian and Lagrangian systems, and second in the appearance of an entropy inequality (1.26) for a linearly degenerate system (1.16). The former phenomenon is an essential feature of the analysis of delta-shocks in chapter 2.

The behavior of an entropy density is an essential feature of the discussion of both delta-shocks and singular shocks.

An alternative to considering the weak convergence of approximate solutions of (1.1-1.4) is to consider a constructive argument for existence of a solution. In the present circumstances, a front tracking algorithm is particularly attractive. Here the term "front-tracking" is used in the sense of Bressan; for background on this method we refer to [Br, BCP, BJ, Ri]. In such an application, one chooses a class A_ε of "acceptable" discontinuities, depending on a parameter ε. The class A_ε must satisfy three properties. First, to each discontinuity in A_ε, there must correspond a speed $s \in \mathbb{R}$ such that the given discontinuity moving with speed s, is a solution (or sufficiently good approximate solution) of (1.1). Second, there must be piecewise constant approximations $w_\varepsilon(\cdot, 0)$ of the initial data, with discontinuities in A_ε such that $w_\varepsilon(\cdot, 0) \to w(\cdot, 0)$ in measure as $\varepsilon \downarrow 0$ with $var(w_\varepsilon(\cdot, 0))$ bounded uniformly in ε. And third, the class A_ε must be complete in the sense that the initial value problem obtained from the collision of any two discontinuities in A_ε must be solvable in the class of piecewise constant functions w with discontinuities corresponding to elements of A_ε.

For obtaining ordinary weak solutions of (1.1), A_ε typically contains entropy shocks, contact discontinuities, and entropy violating shocks (or approximate shocks) of strength not exceeding ε. For the system (1.16), however, the above discussion shows that even for smooth initial data this choice of A_ε will not do, as the required completeness condition is not satisfied.

A larger class A_ε is needed. For the example (1.16), indeed for any system (1.1) with classically unsolvable Riemann problems, a set of jump discontinuities will not suffice for A_ε. If the spatial variation of the elements of A_ε is to remain confined to a single point, distributions must be included. Measures are the simplest of such, and are attractive in the context of maintaining the concept of local conservation of material.

Within the class of measures, the natural candidates for membership in A_ε are of the form

$$w^0(x,t) = \begin{cases} w_-, x - st < 0 \\ w_+, x - st > 0 \end{cases} + \delta(x-st)\big[M_0 + t\big(s(w_+ - w_-) - q(w_+) + q(w_-)\big)\big] \tag{2.7}$$

for some $w_\pm \in D, s \in \mathbb{R}, M_0 \in \mathbb{R}^n$. These are the measures which satisfy a restricted form of (1.1),

$$\lim_{\varepsilon \downarrow 0} \int (w^\varepsilon(x,t)\theta_t(x,t) + q(w^\varepsilon(x,t))\theta_x(x,t))dxdt = 0. \tag{2.8}$$

In (2.8) $\theta \in C^1$ of compact support and satisfies

$$\theta_x(x,t) = 0, |x - st| < \delta_1 \tag{2.9}$$

for some $\delta_1 > 0$. The limit (2.8) holds for any sequence $\{w^\varepsilon\}$ converging in measure to w^0 given by (2.7), as $\varepsilon \downarrow 0$, boundedly in the region $|x - st| > \delta_2$ for some $\delta_2 < \delta_1$.

The set of values of w_+, w_-, s, M_0 in (2.7) for $w^0 \in A_\varepsilon$ is necessarily restricted. One requires that A_ε be complete in the sense described above. The elements of A_ε are often required to satisfy (1.1), or a regularization thereof, in some additional sense. Finally, the set A_ε is required to be invariant under symmetries applicable to the system (1.1).

Indeed, for some systems (1.1) with sufficiently rich symmetry groups, application of a front-tracking algorithm with such an extended set A_ε results in constructive solution of the corresponding Cauchy problem with "large data".

The obtained solutions are valued as n-vector measures on $\mathbb{R}$ at each t, and satisfy (1.1) in the sense of (2.8, 2.9), i.e. with $\theta \in C^1$ and $\theta_x = 0$ in some neighborhood of the singular support of w.

A discussion of the symmetry group of a system (1.1) in this context is given in sections 6 and 7 below. Applications to front-tracking algorithms are given in [KLS] and in section 6 of chapter 3.

A comment on the region D in phase space within which solutions are sought is in order. In any specific example, considerable freedom typically exists in the choice of D. In particular, D does not have to be the largest open subset of $\mathbb{R}^n$ on which q is smoothly defined. Nevertheless, D must contain the values of $w^\varepsilon, \varepsilon > 0$, and still be such that (2.5) holds. Typically D is unbounded; alternatively, D is such that q cannot be continuously extended up to at least some part of the boundary ∂D. With these restrictions, D may be chosen however convenient.

3. Form of distribution solutions

The anticipated form and regularity of distribution solutions of systems of conservation laws is based on experience with familiar "model problems". Perhaps the best-known such systems are zero pressure gas dynamics [B, CLZ, ERS, LW, LZ, TZ, TZZ].

$$\begin{aligned} &\rho_t + (\rho u)_x = 0 \\ &(\rho u)_t + (\rho u^2)_x = 0, \rho > 0, |u| < c, \end{aligned} \tag{3.1}$$

and the "model problem for singular shocks" [KK1, KK2, KK3, KK4, KK5]

$$\begin{aligned} &u_t + (u^2 - v)_x = 0 \\ &v_t + (\frac{u^3}{3} - u)_x = 0, \;\; u, v \in \mathbb{R}. \end{aligned} \tag{3.2}$$

A comment on notation. Throughout u will denote a scalar function of w, possibly one of the components of w, generally interpreted as a fluid velocity, for example in discussions of symmetries. The scalar ρ will always be positive, interpreted as a fluid density. The scalar v will often be interpreted as a specific volume, but less strongly restricted as to assumed values. In (3.1) and below, c is a generic constant distinguished by subscripts as necessary.

The system (3.1) is not hyperbolic; u is a characteristic speed of multiplicity two, with a single corresponding eigenvector [B, ERS, LZ, TZ]. The system admits a (not strictly) convex entropy density U and corresponding flux F.

$$U = \frac{1}{2}\rho u^2, \;\; F = uU. \tag{3.3}$$

The Hugoniot locus for this system contains only two branches

$$\Gamma(\rho_0, u_0) = \{(\rho, u) \neq (\rho_0, u_0) | u = u_0\}. \tag{3.4}$$

The system (3.2) is strictly hyperbolic, with characteristic speeds [KK1, KK2, KK3]

$$\lambda_{\pm}(u, v) = u \pm 1,$$

genuinely nonlinear in the sense of Lax, and admits a strictly convex entropy density

$$U(u, v) = \exp(\frac{1}{2}u^2 - v), \;\; F = uU. \tag{3.5}$$

But the Hugoniot locus for this system is "figure-eight shaped", bounded in particular,

$$\begin{aligned} \Gamma(u_0, u_0) = \{(u, v) \neq (u_0, v_0) | \\ [v - v_0 - \left(\frac{u + u_0}{2}\right)(u - u_0)]^2 + \frac{(u - u_0)^4}{12} = (u - u_0)^2\}. \end{aligned} \tag{3.6}$$

The system (3.1) is related to that describing isothermal, ideal gas dynamics [Ke]

$$\begin{aligned} &\rho_t + (\rho u)_x = 0 \\ &(\rho u)_t + (\rho u^2 + \rho)_x = 0; \end{aligned} \tag{3.7}$$

indeed smooth solutions of (3.7) and (3.2) satisfy each other with the identification

$$v = \frac{1}{2}u^2 - \log \rho; \tag{3.8}$$

using (3.8), (3.5) is recovered as the first equation in (3.7).

Furthermore, the system (3.7) is uniquely solvable in the class of entropy weak solutions, for initial data of large but finite variation and $\rho(\cdot,0)$ uniformly positive [N, Ol]; indeed Euler systems with "nearby" equations of state are also solvable for large variation initial data [T2].

For present purposes we include the strictly hyperbolic system (1.16) as a suitable model problem, with entropy density /flux given in (1.25). For this system, we take

$$D = \{(u,v) | |u| < c,\ 0 < v < c\} \tag{3.9}$$

for some suitably large values of c.

The form of distribution solutions is anticipated by considering entropy inequalities applied to the approximate solutions w^ε. Our model problems all admit convex entropies, scalar functions $U(w)$ satisfying $U_w(w)q_w(w) = F_w(w)$ for all $w \in D$, with F the corresponding entropy flux. The entropy density U cannot be uniformly convex, as this would generally imply an L_2 bound on the $w^\varepsilon(\cdot,t)$ for all $t > 0$, uniformly with respect to ε, and thus preclude weak limits valued as distributions. But in each case the entropy density U satisfies a weaker condition, that for some $a \in \mathbb{R}^n, c > 0$,

$$|w| \le a \cdot w + cU(w) \quad \text{for all } w \in D. \tag{3.10}$$

THEOREM 3.1. *Let $\{w^\varepsilon\}$ be a weakly convergent sequence of approximate solutions of (1.1), satisfying (2.1), (2.2) and (2.4), corresponding to uniformly bounded initial data $w(\cdot,0)$. Assume that the w^ε are bounded uniformly with respect to x,t,ε for x outside a finite interval $J \subset \mathbb{R}$. Assume in addition that there exists an entropy density U satisfying (3.10) and that the w^ε satisfy*

$$U(w^\varepsilon)_t + F(w^\varepsilon)_x \le c \tag{3.11}$$

in the sense of distributions. Then for almost all $\tau > 0$ the weak limit $w^0(\cdot,\tau)$ is a locally bounded measure on $\mathbb{R}$.

REMARK. A simple example illustrates the importance of such an entropy condition (3.11) in obtaining this result. For $1/\varepsilon \in \mathbb{Z}_+$,

$$u^\varepsilon(x,t) = \begin{cases} 0, x < -\varepsilon \\ \pi/\varepsilon^2, -\varepsilon < x < 0 \\ -\pi/\varepsilon^2, 0 < x < \varepsilon \\ 0, x > \varepsilon \end{cases}$$

is a (non-entropy) weak solution of the scalar conservation law

$$u_t + (\sin u)_x = 0$$

with

$$u^\varepsilon \rightharpoonup \pi\delta'(x)$$

weakly as $\varepsilon \downarrow 0$.

PROOF. It suffices to consider the weak limit w^0 within the finite interval J. For any fixed $\tau > 0, \delta > 0$ we choose smooth nonnegative test functions of bounded support for (2.1) and (3.11) of the form

$$\theta(x,t) = \theta^{(1)}(x)\theta^{(2)}(t)$$
$$\theta^{(1)}(x) = 1, x \in J$$
$$\theta^{(2)}(t) = 1, 0 \le t \le \tau$$
$$\theta^{(2)}_t(t) \le 0$$
$$\theta^{(2)}(t) = 0, t \ge \tau + \delta \tag{3.12}$$

thus obtaining from (2.1), after a partial integration in t,

$$-\int_{\tau}^{\tau+\delta} \theta_t^{(2)}(t) \int_J a \cdot w^\varepsilon(x,t)dxdt \le c \tag{3.13}$$

for all sufficiently small ε. Similarly from (3.11) we find

$$-\int_{\tau}^{\tau+\delta} \theta_t^{(2)}(t) \int_J U(w^\varepsilon(x,t))dxdt \le c.$$

Thus from (3.10), for all sufficiently small ε and any such $\theta^{(2)}$

$$\int_{\tau}^{\tau+\delta} |\theta_t^{(2)}(t)| \int_J |w^\varepsilon(x,t)|dxdt \le c; \tag{3.14}$$

given the arbitrariness of $\theta^{(2)}$, it follows that for almost all τ

$$\int_J |w^\varepsilon(x,\tau)|dx \le c \tag{3.15}$$

and thus that the weak limit w^0 is a locally bounded measure on $\mathbb{R}$. □

Much stronger results can be obtained for systems such as (1.16) or (3.1). Indeed, for a class of systems for which (1.16) and (3.1) are examples, we can establish rigorously the general anticipated structure of distribution solutions. For the systems (1.16) and (3.1), $a \cdot w$ satisfying (3.10) is a scalar multiple of $1/v$ or ρ, respectively, and the "fluid velocity" $u(w)$, obtained from

$$a \cdot q(w) = u(w) a \cdot w \tag{3.16}$$

satisfies

$$|u(w)| \le c_0, \quad \text{for all } w \in D. \tag{3.17}$$

Furthermore, using (3.17) we find that for such systems an estimate stronger than (3.10) holds,

$$|w| \le c a \cdot w, \quad \text{for all } w \in D, \tag{3.18}$$

so the singularities in w^0 correspond to those in the single component $a \cdot w^0$.

Finally, an additional entropy condition of the form [Lf, LW, LZ, TZ]

$$u_x \le c_1, \tag{3.19}$$

in the sense of distributions, is typically imposed on the discontinuities in solutions of such systems.

Distribution solutions of (1.1) are anticipated of the form

$$w^0(x,t) = \tilde{w}(x,t) + \sum_i M_i(t)\chi_{I_i}(t)\delta(x - x_i(t)), \tag{3.20}$$

in which $\tilde{w}$ is a weak solution of (1.1) in any open region not containing singularities. In (3.20)

$$\chi_{I_i}(t) = \begin{cases} 1, & t \in I_i \\ 0, & t \notin I_i, \end{cases} \quad \text{with } I_i = [I_i^-, I_i^+) \subset \mathbb{R}_+;$$

$$\begin{array}{rcl} M_i & \in & L_\infty : I_i \to \mathbb{R}^n \\ x_i & \in & W^{1,\infty} : I_i \to \mathbb{R} \end{array}$$

such that $x_i(t) \neq x_j(t)$ for all $i \neq j,\ t \in I_i \cap I_j$.

We can establish the form (3.20) for systems satisfying (3.16-3.18), as the admissible approximate solutions w^ϵ of such systems are readily described using a Lagrangian space variable y.

For each $\epsilon > 0$, assuming

$$a \cdot w^\epsilon(x,t) > 0 \tag{3.21}$$

we find $X^\epsilon : \mathbb{R} \times \mathbb{R}_+ \to \mathbb{R}$ from

$$\begin{aligned} \frac{\partial X^\epsilon(y,t)}{\partial y} &= \frac{1}{a \cdot w^\epsilon(X^\epsilon(y,t),t)} \\ &\stackrel{\text{def}}{=} v^\epsilon(y,t) \end{aligned} \tag{3.22}$$

with $X^\epsilon(0,t)$ a smooth function of t, independent of ϵ.

We denote

$$u^\epsilon(y,t) \stackrel{\text{def}}{=} \frac{\partial X^\epsilon(y,t)}{\partial t}, \tag{3.23}$$

observing from (3.16) that $u^\epsilon(y,t)$ would equal $u(w(X^\epsilon(y,t),t))$ if $a \cdot w_t^\epsilon + a \cdot q(w^\epsilon)_x$ were to vanish identically.

THEOREM 3.2. *Assume a system (1.1) and region $\mathcal{D}$ such that (3.16), (3.18) hold, and a weakly convergent sequence of approximate solutions $\{w^\epsilon\}$, $\epsilon > 0$, satisfying (2.1), (2.2), (2.4). Assume uniformly bounded initial data $w(\cdot,0) = w^\epsilon(\cdot,0)$, for all $\epsilon > 0$, and satisfying*

$$a \cdot w(x,0) \geq c > 0,\ x \in \mathbb{R}. \tag{3.24}$$

Assume further that the approximations w^ϵ satisfy (3.21),

$$|w^\epsilon(x,t)| \leq c,\ x \notin J,\ t > 0,\ \epsilon > 0,\ J \subset \mathbb{R} \text{ a finite interval;} \tag{3.25}$$

$$|u^\epsilon(y,t)| \leq c_0,\ y \in \mathbb{R},\ t > 0,\ \epsilon > 0; \tag{3.26}$$

and in addition

$$u_y^\epsilon(y,t) \leq c_1 v^\epsilon(y,t) \tag{3.27}$$

in any region within which either (3.21) fails to hold uniformly or $|w^\epsilon(X^\epsilon(y,t),t)|$ is not bounded uniformly with respect to y, t, ϵ. Then the weak limit w^0 is of the form (3.20), and in addition

$$|x_{i,t}(t)| \leq c_0,\ t \in I_i,\ i = \cdots. \tag{3.28}$$

REMARKS. With v^ϵ obtained from (3.22), (3.27) is just (3.19) applied to the approximate velocities u^ϵ. The one sided inequalities (3.19) and (3.27) would not be expected in a neighborhood of centered rarefaction waves; thus the mild relaxation in the assumption (3.27). In any event, the characteristic fields for the systems (1.16) and (3.1) are all linearly degenerate, as is typical for systems satisfying (3.16) and (3.18) [S4, Z]. Thus rarefaction waves are not of immediate concern here.

Below M_i will be called the singular mass.

PROOF. In any region where (3.21) holds, uniformly, v^ϵ is uniformly bounded from (3.22). Alternatively, in any region where (3.27) holds, from (3.22), (3.23), (3.27)

$$\begin{aligned} v_t^\epsilon(y,t) &= u_y^\epsilon(y,t) \\ &\leq c_1 v^\epsilon(y,t) \end{aligned}$$

and thus

(3.29) $$v^\varepsilon(y,t) \leq ce^{c_1 t} \quad \text{for all} \quad y \in \mathbb{R}, \varepsilon > 0.$$

Therefore for any $\tau > 0$ we have both first partial derivatives of X^ε uniformly bounded on $\mathbb{R}\times(0,\tau)$ uniformly with respect to ε. Taking a subsequence if necessary as $\varepsilon \downarrow 0$, we find

(3.30) $$X^\varepsilon \to X^0$$

uniformly on any compact subset of $\mathbb{R} \times [0,\tau]$,

(3.31) $$X_t^\varepsilon \rightharpoonup X_t^0, X_y^\varepsilon \rightharpoonup X_y^0$$

weakly on $\mathbb{R} \times [0,\tau]$. Thus

(3.32) $$X^0 \in W^{1,\infty} : \mathbb{R} \times [0,\tau] \to \mathbb{R}$$

with

(3.33) $$|X_t^0| \leq c_0$$

(3.34) $$0 \leq X_y^0 \leq ce^{c_1\tau}.$$

Next we show that the entire sequence converges. From (3.22),

(3.35) $$a \cdot w^\varepsilon(X^\varepsilon(y,t),t)\frac{\partial X^\varepsilon}{\partial y}(y,t) = 1,$$

and thus for any $\theta \in C_0 : \mathbb{R} \to \mathbb{R}$

(3.36) $$\int_{\mathbb{R}} \theta(x)a \cdot w^\varepsilon(x,t)dx = \int_{\mathbb{R}} \theta(X^\varepsilon(y,t))dy$$

pointwise in t. As $\varepsilon \downarrow 0$, the left side of (3.36) converges (no subsequence necessary), and thus the limit X^0 in (3.30) is unique and no subsequence is necessary. In addition, we have from (3.30) and (3.36)

$$\int_{\mathbb{R}} \theta(x)a \cdot w^0(x,t)dx = \int_{\mathbb{R}} \theta(X^0(y,t))dy$$

pointwise in t, and thus

(3.37) $$a \cdot w^0(X^0(y,t),t) = 1/\frac{\partial X^0}{\partial y}(y,t)$$

whenever the right side of (3.37) is defined.

Next for $y_1 > y_0, \varepsilon > 0$,

$$\begin{aligned}\frac{d}{dt}(X^\varepsilon(y_1,t) - X^\varepsilon(y_0,t)) &= u^\varepsilon(X^\varepsilon(y_1,t),t) - u^\varepsilon(X^\varepsilon(y_0,t),t) \\ &= \int_{X^\varepsilon(y_0,t)}^{X^\varepsilon(y_1,t)} u_x(x,t)dx \\ &\le c(X^\varepsilon(y_1,t) - X^\varepsilon(y_0,t)) \end{aligned} \tag{3.38}$$

with c independent of using (3.19) and (3.26). Thus such differences $X^\varepsilon(y_1,t) - X^\varepsilon(y_0,t)$ can increase at most exponentially with respect to time, uniformly with respect to ε, and we have shown that for $t_1 > t_0,\ t_0, y_1, y_0$ such that

$$\begin{aligned} X^0(y_1,t_0) - X^0(y_0,t_0) &= \lim_{\varepsilon\downarrow 0} X^\varepsilon(y_1,t_0) - X^\varepsilon(y_0,t_0) \\ &= 0, \end{aligned} \tag{3.39}$$

then

$$\begin{aligned} X^0(y_1,t_1) - X^0(y_0,t_1) &= \lim_{\varepsilon\downarrow 0} X^\varepsilon(y_1,t_1) - X^\varepsilon(y_0,t_1) \\ &= 0. \end{aligned} \tag{3.40}$$

Thus the intervals with respect to y within which $X^0(y,t)$ is independent of y are noncontracting on both ends as t increases. This is the familiar "adhesion dynamics" or "sticky particle" condition well-known in the theory of delta-shocks [B, BG, Lf, LW].

We identify each such disjoint interval with one of the $x_i(t)$, the constant value of $X^0(y,t)$ within such an interval. From (3.33) this can be done so that (3.24) is satisfied and so that

$$X^0(y,t) \in \Omega^t \stackrel{\text{def}}{=} \bigcup_{i:t\in I_i} x_i(t) \tag{3.41}$$

for almost all (y,t) such that $X^0_y(y,t) = 0$.

It remains to recover $\tilde{w}$ and M_i so that (3.20), (3.21) hold.

For $\theta \in C_0 : J \to \mathbb{R}^n, t > 0$ fixed,

$$\begin{aligned}\int_{\mathbb{R}} \theta(x)\cdot w^0(x,t)dx &= \lim_{\varepsilon\downarrow 0}\int_{\mathbb{R}} \theta(x)\cdot w^\varepsilon(x,t)dx \\ &= \lim_{\varepsilon\downarrow 0}\int_{\mathbb{R}} \theta(X^\varepsilon(y,t))\cdot w^\varepsilon(X^\varepsilon(y,t),t)X^\varepsilon_y(y,t)dy \\ &= \lim_{\varepsilon\downarrow 0}\int_{\mathbb{R}} \frac{\theta(X^\varepsilon(y,t))\cdot w^\varepsilon(X^\varepsilon(y,t),t)}{a\cdot w^\varepsilon(X^\varepsilon(y,t),t)}dy \end{aligned} \tag{3.42}$$

using (3.35). The integral in (3.42) is partitioned

$$\int_{\mathbb{R}} = \int_{y:X^0(y,t)\notin\Omega^t} + \int_{y:X^0(y,t)\in\Omega^t} . \tag{3.43}$$

For $\delta > 0$ set

$$\theta^\delta(x) = \begin{cases} \theta(x), & \operatorname{dist}(x, \Omega^t) > \delta \\ 0, & \operatorname{dist}(x, \Omega^t) \leq \delta. \end{cases} \tag{3.44}$$

From (3.18), the integrand in (3.42) is uniformly bounded. Thus the first integral in (3.43) is

$$\begin{aligned} &\lim_{\varepsilon\downarrow 0}\lim_{\delta\downarrow 0} \int\limits_{y:X^0(y,t)\notin\Omega^t} \frac{\theta^\delta(X^\varepsilon(y,t))\cdot w^\varepsilon(X^\varepsilon(y,t),t)}{a\cdot w^\varepsilon(X^\varepsilon(y,\varepsilon),t)}dy \\ &= \lim_{\delta\downarrow 0}\lim_{\varepsilon\downarrow 0} \int\limits_{y:X^0(y,t)\notin\Omega^t} \frac{\theta^\delta(X^\varepsilon(y,t))\cdot w^\varepsilon(X^\varepsilon(y,t),t)}{a\cdot w^\varepsilon(X^\varepsilon(y,t),t)}dy \end{aligned} \tag{3.45}$$

the interchange of limits justified by (3.18) and (3.44).

Since $X^\varepsilon(\cdot,t)\to X^0(\cdot,t)$ uniformly as $\varepsilon\downarrow 0$, for ε sufficiently small the support of $\theta^\delta(X^\varepsilon(\cdot,t))$ does not intersect Ω^t from (3.44), and so using (3.35) again (3.45) becomes

$$\begin{aligned} &\lim_{\delta\downarrow 0}\lim_{\varepsilon\downarrow 0} \int\limits_{y:X^0(y,t)\notin\Omega^t} \theta^\delta(X^\varepsilon(y,t))\cdot w^\varepsilon(X^\varepsilon(y,t),t)X_y^\varepsilon(y,t)dy \\ &= \lim_{\delta\downarrow 0}\lim_{\varepsilon\downarrow 0} \int\limits_{\mathbb{R}} \theta^\delta(x)\cdot w^\varepsilon(x,t)dx \\ &= \lim_{\delta\downarrow 0} \int\limits_{\mathbb{R}} \theta^\delta(x)\cdot w^0(x,t)dx \\ &\stackrel{\text{def}}{=} \int\limits_{\mathbb{R}} \theta(x)\tilde{w}(x,t)dx \end{aligned} \tag{3.46}$$

so $\tilde{w}(\cdot,t)$ is just $w^0(\cdot,t)$ without the singularities on Ω^t, as expected. Since $w^0(\cdot,t)$ is a locally bounded measure on $\mathbb{R}$, $\tilde{w}(\cdot,t)\in L_1:\mathbb{R}\to\mathbb{R}^n$ and (3.21) holds.

For the second integral in (3.43), we use the continuity of θ, the uniform convergence of $X^\varepsilon(\cdot,t)$ to $X^0(\cdot,t)$ and (3.18) to obtain

$$\begin{aligned} &\lim_{\varepsilon\downarrow 0} \int\limits_{y:X^0(y,t)\in\Omega^t} \frac{\theta(X^\varepsilon(y,t))\cdot w^\varepsilon(X^\varepsilon(y,t),t)}{a\cdot w^\varepsilon(X^\varepsilon(y,t),t)}dy \\ &= \lim_{\varepsilon\downarrow 0} \int\limits_{y:X^0(y,t)\in\Omega^t} \frac{\theta(X^0(y,t))\cdot w^\varepsilon(X^\varepsilon(y,t),t)}{a\cdot w^\varepsilon(X^\varepsilon(y,t),t)}dy \\ &= \lim_{\varepsilon\downarrow 0} \sum_{i:t\in I_i} \theta(x_i(t))\cdot \int\limits_{y:X^0(y,t)=x^i(t)} \frac{w^\varepsilon(X^\varepsilon(y,t),t)}{a\cdot w^\varepsilon(X^\varepsilon(y,t),t)}dy \end{aligned} \tag{3.47}$$

so (3.20) holds with

$$M_i(t) = \lim_{\varepsilon\downarrow 0} \int\limits_{y:X^0(y,t)=x_i(t)} \frac{w^\varepsilon(X^\varepsilon(y,t),t)}{a\cdot w^\varepsilon(X^\varepsilon(y,t),t)}dy,\ i=1,\dots. \tag{3.48}$$

The proof is complete. □

Some comments regarding the singular mass M_i in (3.48) are in order. For i fixed, $t \in I_i$, let $[y_-, y_+]$ be the maximal interval such that

$$X^0(y,t) = x_i(t), y \in [y_-, y_+]; \tag{3.49}$$

then from (3.48)

$$a \cdot M_i(t) = y_+ - y_-. \tag{3.50}$$

Using (3.35) in (3.48) we find

$$M_i(t) = \lim_{\varepsilon \downarrow 0} \int_{X^\varepsilon(y_-,t)}^{X^\varepsilon(y_+,t)} w^\varepsilon(x,t)dx \tag{3.51}$$

so for nonzero $M_i(t), |w^\varepsilon(\cdot,t)|$ cannot be bounded in a neighborhood of $x_i(t)$ uniformly in ε.

The integrand in (3.48) is uniformly bounded by (3.18); should it be continuous in an open neighborhood of $[y_-, y_+]$ uniformly with respect to ε, one may replace X^ε with X^0 and obtain from (3.49)

$$M_i(t) = (y_+ - y_-) \lim_{\varepsilon \downarrow 0} \frac{w^\varepsilon(x_i(t),t)}{a \cdot w^\varepsilon(x_i(t),t)}. \tag{3.52}$$

Such continuity uniformly in ε in however exceptional.

Whether or not the conditions of theorem 3.2 are satisfied, with fairly mild restrictions on the w^ε the form (3.20) of the weak limit w^0 implies important conditions on $\tilde{w}$ and M_i, and on the weak limit q^0 of $q(w^\varepsilon)$, obtained in (2.5) and satisfying (2.6).

THEOREM 3.3. *Assume that the w^ε satisfy (2.1) and that as $\varepsilon \downarrow 0$,*

$$w^\varepsilon(\cdot,t) \to w^0(\cdot,t) \quad \textit{weakly in the space of measures on } \mathbb{R} \tag{3.53}$$

pointwise in t, with w^0 of the form (3.20), satisfying (3.24), and

$$w^\varepsilon \to \tilde{w} \quad \textit{in measure on } \mathbb{R} \times [0,\tau] \quad \textit{for any fixed } \tau > 0. \tag{3.54}$$

Assume in addition that the w^ε are uniformly bounded away from the singularities, i.e. that for any $\delta > 0$ there exists c_δ such that

$$|w^\varepsilon(x,t)| \le c_\delta \quad \textit{uniformly with respect to } \varepsilon, \tag{3.55}$$

for all (x,t) satisfying

$$dist((x,t), \bigcup_i \{x_i(I_i)\}) \ge \delta. \tag{3.56}$$

Then the weak limit q^0 (from (2.5), (2.6)) is of the form

$$q^0(x,t) = q(\tilde{w}(x,t)) + \sum_i M_i(t) x_{i,t}(t) \chi_{I_i}(t) \delta(x - x_i(t), \tag{3.57}$$

and

$$\tilde{w}_t + q(\tilde{w})_x = 0 \tag{3.58}$$

weakly in any open region of $\mathbb{R} \times \mathbb{R}_+$ not containing singularities. Furthermore, each singular mass satisfies

$$\begin{aligned} \frac{dM_i(t)}{dt} &= x_{i,t}(t)\Big(\tilde{w}(x_i(t)+0,t) - \tilde{w}(x_i(t)-0,t)\Big) \\ &\quad - \Big(q(\tilde{w}(x_i(t)+0,t)) - q(\tilde{w}(x_i(t)-0,t))\Big) \end{aligned} \tag{3.59}$$

weakly in I_i. Finally, at a point $(\hat{x},\hat{t})$ where singularities collide, singular mass is conserved

$$\sum_{i:x_i(\hat{t}-0)=\hat{x}} M_i(\hat{t}-0) = \sum_{j:x_j(\hat{t}+0)=\hat{x}} M_j(\hat{t}+0). \tag{3.60}$$

PROOF. In any compact subset of $\mathbb{R} \times \mathbb{R}_+$ not containing singularities, w^ε converges boundedly in measure to $\tilde{w}$ from (3.55), (3.56), so (3.58) follows from (2.6), and q^0 is of the form

$$q^0 = q(\tilde{w}) + \zeta \tag{3.61}$$

where ζ is a distribution with support confined to the singularities.

We apply (2.6) with a test function $\theta \in C_0^\infty : \mathbb{R} \times \mathbb{R}_+ \to \mathbb{R}^n$ with support containing one such singularity, obtaining w^0 from (3.20) and q^0 from (3.61); after the usual partial integrations

$$\begin{aligned} &\int_{\mathbb{R}_+}\int_{\mathbb{R}} (\tilde{w}\cdot\theta_t + q(\tilde{w})\cdot\theta_x)dxdt + \int_{I_i} M_i(t)\cdot\theta_t(x_i(t),t)dt \\ &\quad + \int_{I_i} \zeta(x_i(t),t)\cdot\theta_x(x_i(t),t)dt = 0 \end{aligned} \tag{3.62}$$

In view of (3.58), the first term in (3.62) depends only on the possible discontinuity of $\tilde{w}$ on the singularity, and is equal to

$$\begin{aligned} &\int_{I_i} \theta(x_i(t),t)\cdot[-\tilde{w}(x_i(t),t+0) + \tilde{w}(x_i(t),t-0) \\ &\qquad - q(\tilde{w}(x_i(t)+0,t)) + q(\tilde{w}(x_i(t)-0,t))]dt \\ &= \int_{I_i} \theta(x_i(t),t)\cdot[x_{i,t}(t)(\tilde{w}(x_i(t)+0,t) - \tilde{w}(x_i(t)-0,t)) \\ &\qquad - q(\tilde{w}(x_i(t)+0,t)) + q(\tilde{w}(x_i(t)-0,t))]dt. \end{aligned} \tag{3.63}$$

In the second term in (3.62) we use

$$\theta_t(x_i(t),t) = \frac{d}{dt}\theta(x_i(t),t) - x_{i,t}(t)\theta_x(x_i(t),t), \tag{3.64}$$

obtaining after a partial integration

$$\begin{aligned} \int_{I_i} M_i(t)\cdot\theta_t(x_i(t),t)dt &= -\int_{I_i}\Big[\frac{dM_i(t)}{dt}\cdot\theta(x_i(t),t) \\ &\qquad + x_{i,t}(t)M_i(t)\cdot\theta_x(x_i(t),t)\Big]dt \end{aligned} \tag{3.65}$$

Combining (3.62), (3.63) and (3.64), we find

$$\int_{I_i} \theta(x_i(t),t) \cdot \Big[x_{i,t}(t)(\tilde{w}(x_i(t)+0,t) - \tilde{w}(x_i(t)-0,t))$$
$$- q(\tilde{w}(x_i(t)+0,t)) + q(\tilde{w}(x_i(t)-0,t)) - \frac{dM_i(t)}{dt}\Big] dt$$
$$+ \int_{I_i} \theta_x(x_i(t),t) \cdot \Big[\zeta(x_i(t),t) - x_{i,t}(t) M_i(t)\Big] dt = 0. \tag{3.66}$$

Since the values of θ and θ_x on the singularities are arbitrary and independent, the two terms in brackets must vanish. The vanishing of the first such expression establishes (3.59), and that of the second establishes (3.57), using (3.61).

Thus it remains only to obtain (3.60). Consider (2.6) with a smooth test function θ the support of which contains a single collision point $(\hat{x}, \hat{t})$. Using (3.20) for w^0 and (3.57) for q^0, in view of (3.58) and (3.62) we have

$$0 = \int\int (w^0 \cdot \theta_t + q^0 \cdot \theta_x) dx dt$$
$$= - \int_{\mathbb{R}_+} \sum_i \theta(x_i(t),t) \cdot M_i(t) \frac{d}{dt} \chi_{I_i}(t) dt \tag{3.67}$$

from which (3.60) follows easily. □

Theorem 3.3 has been obtained under the tacit assumption that the limiting values $\tilde{w}(x_i(t) \pm 0, t)$ exist for almost all t. This is unnecessary; having established that the 1-form $\tilde{w}dx - q(\tilde{w})dt$ is weakly closed except on the singularities, one recovers (3.57), (3.60) and obtains a more general form of (3.59) by seeking a closed 1-form of the form

$$\Xi(x,t) = \tilde{w}(x,t)dx - q(\tilde{w}(x,t))dt$$
$$+ \sum_i M_i(t) \chi_{I_i}(t) \delta(x - x_i(t)) dx - \zeta dt \tag{3.68}$$

by the same argument. Again the support of ζ is confined to the singularities.

The equation (3.59), however, is the familiar "generalized Rankine-Hugoniot relation" in the theory of delta-shocks [LZ]. This equation provides n relations among the $3n+1$ variables $\tilde{w}(x_i(t) \pm 0, t), M_i(t)$, and $x_{i,t}(t)$. Thus even if the singularity is completely overcompressive, with all n characteristics on both sides approaching the singularity with increasing time, one additional relation is needed to determine these variables uniquely.

Indeed, almost all examples of singularities in distribution solutions of systems of conservation laws are completely overcompressive. The additional required relation between $\tilde{w}(x_i(t) \pm 0, t), M_i(t)$, and $x_{i,t}(t)$ is obtained for a given system (1.1) by identifying $\mathcal{M} \subset \mathbb{R}^n$, the set of all possible values of singular mass $M_i(t)$. From an elementary symmetry argument, given in section 6 below, it follows that $\mathcal{M}$ is a cone. In the case $\dim \mathcal{M} < n$, there is a nontrivial subspace $\mathcal{M}_\perp \perp \mathcal{M}$, so using $\mathcal{M}_\perp \perp dM_i(t)/dt$ in (3.59) we obtain $n - \dim \mathcal{M}$ additional relations satisfied by the singularity. With $n - \dim \mathcal{M}$ of the classical Rankine-Hugoniot relations thus satisfied, $\tilde{w}(x_i(t) \pm 0, t)M_i(t)$ and $x_{i,t}(t)$ can be uniquely determined for a singularity with $n + 1 + \dim \mathcal{M}$ incoming characteristics and $n - 1 - \dim \mathcal{M}$ outgoing characteristics. In almost all examples $\dim \mathcal{M} = n - 1$.

In the case $\dim \mathcal{M} = n$, for $M_i(t)$ in the interior of $\mathcal{M}$, (3.59) is solved for $dM_i(t)/dt$ for any given values of $\tilde{w}(x_i(t) \pm 0, t), x_{i,t}(t)$. In this case $x_{i,t}(t)$ is determined as a function of $M_i(t)$ from consideration of the form of the given function $q(w)$, and all singularities are completely overcompressive.

It is at this point where a unified treatment of distribution solutions of (1.1) can no longer be continued; a weaker treatment, admitting the dichotomy between delta-shocks and singular shocks, becomes essential.

4. Delta-shocks and singular shocks

The analysis of delta-shocks has little in common with that of singular shocks, the common representation (3.20-3.24), (3.57), (3.59), (3.60) notwithstanding. We distinguish the two cases as follows: In view of the results of section 1.3, we assume a sequence $\{w^\varepsilon\}$ satisfying (2.1) such that pointwise in t, the $w^\varepsilon(\cdot, t)$ are locally bounded measures uniformly in ε, and converge weakly in the space of measures on $\mathbb{R}$ as $\varepsilon \downarrow 0$ to a limit $w^0(\cdot, t)$ of the form (3.20).

DEFINITION. **If the $q(w^\varepsilon(\cdot, t))$ are also locally bounded measures, uniformly with respect to ε, in some open subset of $\mathbb{R} \times \mathbb{R}_+$ containing $\{(x_i(t), t), t \in I_i\}$ then the singularity on x_i is called a delta-shock; otherwise it is called a singular shock.**

Since the w^ε may presumably assume any value in D, it is no surprise that all of the known examples of systems (1.1) admitting delta-shock solutions satisfy a condition

$$|q(w)| \leq c(1 + |w|) \quad \text{for all } w \in D \tag{4.1}$$

with a constant in (4.1) depending on D. For example, for the system (1.16) with D given in (3.9), (4.1) follows easily from (3.9).

For such a sequence $\{w^\varepsilon\}$ and a system satisfying (4.1), it follows that

$$q(w^\varepsilon(\cdot, t)) \rightharpoonup q^0(\cdot, t) \quad \text{as } \varepsilon \downarrow 0, \tag{4.2}$$

weakly in the space of measures on $\mathbb{R}$, pointwise in t, with q^0 of the form (3.57), and

$$q(\tilde{w}(\cdot, t)) \in L_1^{\text{loc}}(\mathbb{R})^n, \quad \text{pointwise in } t. \tag{4.3}$$

An extension of the definition of a weak solution of (1.1) is readily available for systems satisfying (4.1) and (3.18), which includes the known examples of systems admitting delta-shock solutions.

Denote by

$$\Sigma = \{(\frac{1}{a \cdot w}, \frac{w}{a \cdot w}, \frac{q(w)}{a \cdot w}), w \in D\}. \tag{4.4}$$

For some fixed $\tau > 0$, suppose that for each $t \in (0, \tau)$

$$v(\cdot, t) \in L_1^{\text{loc}}(\mathbb{R} \to \mathbb{R}_+), 1/v(\cdot, t) \in L_1^{\text{loc}}, \int_{-\infty}^{\infty} v(y, t) dy = \infty$$

$$\hat{w}(\cdot, t), \hat{q}(\cdot, t) \in L_\infty(\mathbb{R} \to \mathbb{R}^n). \tag{4.5}$$

Then identifying

$$x(y,t) = \int^{y} v(y',t)dy' \tag{4.6}$$

$$w(x,t) = \hat{w}(y,t)/v(y,t), \tag{4.7}$$

from $(v(y,t), \hat{w}(y,t), \hat{q}(y,t)) \in \Sigma$ it follows that $q(w(x,t)) = \hat{q}(y,t)/v(y,t)$. Thus w is a weak solution of (1.1) in the dual of the space of $\theta \in W^{1,\infty}$ of bounded support in $\mathbb{R} \times (0,\tau)$, if and only if

$$(v(y,t), \hat{w}(y,t), \hat{q}(y,t)) \in \Sigma \quad \text{for almost all } y, \text{ pointwise in } t \tag{4.8}$$

and for all such θ

$$\iint (\hat{w}(y,t) \cdot \theta_t(x(y,t),t) + \hat{q}(y,t) \cdot \theta_x(x(y,t),t))dydt = 0. \tag{4.9}$$

The lower limit in the integral in (4.6) is an arbitrary smooth function of t. Different choices thereof would imply different values $x(y,t), x^*(y,t)$ satisfying $x(y,t) - x^*(y,t) = \chi(t)$ for some smooth function χ. Then if $\hat{w}, \hat{q}$ satisfy (4.9), they also satisfy (4.9) with $\theta(x(y,t),t)$ replaced by $\phi(x^*(y,t),t)$ where ϕ determined from $\phi(x^*,t) = \theta(x^* - \chi(t),t)$ is also an eligible test function. Therefore the choice of the lower limit in the integral in (4.6) is arbitrary.

The desired extension of the concept of weak solution is accomplished by allowing v to be zero on a set of finite measure. Denote by

$$\Sigma_0 = \partial\Sigma \cap \{v = 0\} \quad \text{and } \bar{\Sigma} = \Sigma \cup \Sigma_0 \tag{4.10}$$

and suppose that for each $t \in (0,\tau)$

$$v(\cdot,t) \in L_1^{\text{loc}}(\mathbb{R} \to \mathbb{R}_+), \int_{-\infty}^{\infty} v(y,t)dy = \infty$$

$$\hat{w}(\cdot,t), \hat{q}(\cdot,t) \in L_\infty : \mathbb{R} \to \mathbb{R}^n \tag{4.11}$$

Definition. **For a system (1.1) satisfying (4.1) and (3.18),** $(v, \hat{w}, \hat{q})$ **is a distribution solution in** $\mathbb{R} \times (0,\tau)$ **if**

$$(v(y,t), \hat{w}(y,t), \hat{q}(y,t)) \in \bar{\Sigma} \textbf{ for almost all } y, \textbf{ pointwise in } t \in (0,\tau) \tag{4.12}$$

and (4.9) holds for all $\theta \in C_0^1(\mathbb{R} \times (0,\tau) \to \mathbb{R}^n)$ **with** $x(y,t)$ **obtained from (4.6).**

It is clear that if $v, \hat{w}, \hat{q}$ is the limit of a sequence $v^\varepsilon, \hat{w}^\varepsilon, \hat{q}^\varepsilon$ assuming values almost everywhere in Σ and converging boundedly in measure in $\mathbb{R}$, pointwise in t, as $\varepsilon \downarrow 0$, then $v, \hat{w}, \hat{q}$ is a distribution solution if (4.9) holds.

Similarly, a distribution solution $v, \hat{w}, \hat{q}$ is the limit of approximate solutions. Taking $v^\varepsilon(y,t) = v(y,t) + \varepsilon$, there exist $\hat{w}^\varepsilon(y,t), \hat{q}^\varepsilon(y,t)$ such that $(v^\varepsilon(y,t), \hat{w}^\varepsilon(y,t), \hat{q}^\varepsilon(y,t)) \in \Sigma$ and $\hat{w}^\varepsilon(y,t) \to \hat{w}(y,t), \hat{q}^\varepsilon(y,t) \to \hat{q}(y,t)$ as $\varepsilon \to 0$ for almost all y,t. Indeed, one may choose $\hat{w}^\epsilon(y,t) = \hat{w}(y,t),\ \hat{q}^\epsilon(y,t) = v^\epsilon(y,t)q(\hat{w}(y,t)/v^\epsilon(y,t))$. Then determining $x^\varepsilon(y,t)$ as in (4.6), $x^\varepsilon, \hat{w}^\varepsilon, \hat{q}^\varepsilon$ satisfy (4.9) in the limit $\varepsilon \downarrow 0$, so the corresponding w^ε obtained from (4.7) satisfies (1.1) weakly in the limit $\varepsilon \downarrow 0$. A straightforward computation justifying this claim is omitted in the interest of brevity.

Wherever v is nonzero, w and $q(w)$ are obtained from (4.7). More generally, given $(v, \hat{w}, \hat{q})$, $x(y,t)$ is determined from (4.6) and $w, q(w)$ as measures from

$$\int_{\mathbb{R}} w(x,t) \cdot \theta(x) dx = \int_{\mathbb{R}} \hat{w}(y,t) \cdot \theta(x(y,t)) dy$$

$$\int_{\mathbb{R}} q(w(x,t)) \cdot \theta(x) dx = \int_{\mathbb{R}} \hat{q}(y,t) \cdot \theta(x(y,t)) dy$$

for $\theta \in C_0(\mathbb{R} \to \mathbb{R}^n)$.

DEFINITION. **A distribution solution $(v, \hat{w}, \hat{q})$ of a system (1.1) satisfying (4.1) and (3.18) is a delta-shock solution if w is of the form (3.20) and $q(w)$ of the corresponding form (3.57).**

Thus distribution solutions which are weak limits of approximating sequences w^ε in the sense of theorem 3.3 are delta-shock solutions. However, the existence of delta-shock solutions implies restrictions on q and on the set Σ_0 in particular.

For a singularity at some $x_i(t)$ with nonzero singular mass $M_i(t)$ and such that $x_{i,t}(t)$ exists, let (y_-, y_+) be the maximal interval within which $v(\cdot, t)$ vanishes and such that $x(y,t) = x_i(t)$. Then

$$\int_{y_-}^{y_+} \hat{w}(y,t) dy = M_i(t) \tag{4.13}$$

and if this singularity is a delta-shock, satisfying (3.57), then

$$\int_{y_-}^{y_+} \hat{q}(y,t) dy = x_{i,t}(t) M_i(t) \tag{4.14}$$

with the points

$$(0, \hat{w}(y,t), \hat{q}(y,t)) \in \Sigma_0 \quad \text{a.e. in } (y_-, y_+). \tag{4.15}$$

For $\hat{w}(\cdot, t)$ independent of y in this interval, $\hat{q}(\cdot, t)$ will also be independent of y, from (4.4) and it follows from (4.13), (4.14), (3.50) that

$$(0, \frac{M_i(t)}{a \cdot M_i(t)}, x_{i,t}(t) \frac{M_i(t)}{a \cdot M_i(t)}) \in \Sigma_0. \tag{4.16}$$

DEFINITION. **A delta-shock solution of a system (1.1) satisfying (3.18) is single-valued if (4.16) holds for all $M_i(t) \neq 0, x_{i,t}(t)$.**

Systems admitting single-valued delta-shock solutions are generally of the form

$$q(w) = u(\frac{w}{a \cdot w}) w + b(w) \tag{4.17}$$

with u a bounded scalar function of $w \in D$, D a cone, and b satisfying

$$\lim_{\substack{|w| \to \infty \\ w \in D}} |b(w)| / |w| = 0. \tag{4.18}$$

Then from (4.4), (4.10), (4.17)

$$\Sigma_0 = \{(0, \frac{w}{a \cdot w}, u(\frac{w}{a \cdot w}) \frac{w}{a \cdot w}), w \in D\} \tag{4.19}$$

and comparing (4.16) with (4.19) we infer that for such systems $\mathcal{M}$ is isomorphic to D, i.e. $\dim \mathcal{M} = n$. In addition, we obtain the needed additional relation among $x_{i,t}, M_i$ and the limiting values of $\tilde{w}$, of the form

$$x_{i,t}(t) = u(\frac{M_i(t)}{a \cdot M_i(t)}). \tag{4.20}$$

In the case of singular shocks, the form of $\mathcal{M}$ in general depends on properties of the entropy density, as discussed in section 1 of chapter 3. The value of $\dim \mathcal{M}$ and a subspace containing $\mathcal{M}$ may also be apparent from the symmetry group of a given system, as discussed in section 6 below.

However, this information is often available by inspection of a given system. For example, for the model problem (3.2), from (3.51) with $\varepsilon, \delta > 0$

$$M_i^{(1)}(t) = \int_{x_i(t)-\delta}^{x_i(t)+\delta} u^{\varepsilon}(x,t)dx + o(1) \quad \text{as } \varepsilon, \delta \downarrow 0 \tag{4.21}$$

$$M_i^{(2)}(t) = \int_{x_i(t)-\delta}^{x_i(t)+\delta} v^{\varepsilon}(x,t)dx + o(1). \tag{4.22}$$

From (3.57) applied to this system

$$x_{i,t}(t) M_i^{(1)}(t) = \int_{x_i(t)-\delta}^{x_i(t)+\delta} (u^{\varepsilon}(x,t)^2 - v^{\varepsilon}(x,t))dx + o(1) \tag{4.23}$$

and

$$x_{i,t}(t) M_i^{(2)}(t) = \int_{x_i(t)-\delta}^{x_i(t)+\delta} (\frac{1}{3}u^{\varepsilon}(x,t)^3 - u^{\varepsilon}(x,t))dx + o(1). \tag{4.24}$$

Using the Schwarz inequality in (4.13)

$$\begin{aligned} |M_i^{(1)}(t)| &\leq (2\delta)^{1/2} (\int_{x_i(t)-\delta}^{x_i(t)+\delta} u^{\varepsilon}(x,t)^2 dx)^{1/2} + o(1) \\ &= (2\delta)^{1/2} (M_i^{(2)}(t) + x_{i,t}(t) M^{(1)}(t))^{1/2} + o(1) \end{aligned} \tag{4.25}$$

using (4.22) and (4.23). Since (4.25) holds for any $\delta > 0$, $M_i^{(1)}$ vanishes identically (and $M_i^{(2)}$ is nonnegative). Thus from (3.59), the first component of the standard Rankine-Hugoniot relation must be satisfied, providing the needed relation between $x_{i,t}$ and the limiting values of $\tilde{w}$.

For the system (3.2) this is

$$\begin{aligned} &x_{i,t}(t)(u(x_i(t)+0) - u(x_i(t)-0)) \\ &= u(x_i(t)+0)^2 - v(x_i(t)+0) - u(x_i(t)-0)^2 + v(x_i(t)-0). \end{aligned} \tag{4.26}$$

The different form of (4.20) and (4.26) is apparent. For such delta-shock systems, $x_{i,t}$ depends only on the singular mass $M_i(t)$. For singular shock systems such as (3.2), $x_{i,t}$ depends only on the limiting values of $\tilde{w}$. Such is generally true in both cases.

Regularization of (1.1) is an alternative to such an extension of the concept of a weak solution. Indeed, all known examples of singular shocks, with the above definition, correspond to limits of viscous structures [KK3, S5, SSS].

The dichotomy between delta-shocks and singular shocks may be explained by considering the limits

$$\lim_{\varepsilon\downarrow 0}\iint w_t^\varepsilon(x,t)\cdot\theta(x,t)dxdt,\ \lim_{\varepsilon\downarrow 0}\iint q(w^\varepsilon)_x(x,t)\cdot\theta(x,t)dxdt,$$
$$\lim_{\varepsilon\downarrow 0}\epsilon\iint w_x^\varepsilon(x,t)\cdot\theta_x(x,t)dxdt \tag{4.27}$$

given the above distinction between delta-shocks and singular shocks.

In the delta-shock case, the first two limits exist for $\theta\in C_0^1$, but the third requires $\theta\in C_0^2$. This suggests an extension of weak solution based on C_0^1 test functions, but shows that such a regularization could not lead to a sequence of approximations satisfying (2.1) with less smooth test functions. Indeed, for systems satisfying (3.18), results on viscous structure for delta-shocks are quite limited, and based on "incomplete regularization", i.e. using a singular viscosity matrix [CL2, ERS, KK6, Ko, Ob, TZZ, YL].

For singular shocks, in contrast, the first limit requires only $\theta\in C_0^1$, but the second and third limits require smoother θ. One can hope that these two terms will compensate each other in this respect - this provides a point of departure for the discussion of singular shocks in chapter 3.

There exists a simple class of systems (1.1) combining features generally identified with delta-shocks and those generally identified with singular shocks. These are pairs of the form [Ko, CL2, ERS, Ob, TZZ]

$$\begin{aligned}\rho_t+(\rho u)_x&=0\\ g(u)_t+f(u)_x&=0,\end{aligned} \tag{4.28}$$

for which we take

$$D=\{\rho>0,|u|<c\}. \tag{4.29}$$

In (4.28), f,g are smooth functions, with $g(0)=0$ and $g_u(u)>0$; typically $g(u)=u$.

These systems satisfy (3.10) with $U(w)$ depending only on u and $a=\begin{pmatrix}1\\0\end{pmatrix}$. The notation in (4.28) is such that (3.16), (3.17) hold.

These systems do not satisfy (3.18), but do satisfy a weakened form thereof

$$\limsup_{\substack{|w|\to\infty\\ w\in D}}\frac{|w|}{a\cdot w}<\infty. \tag{4.30}$$

Using (4.30), the sets $\Sigma,\Sigma_0,\bar{\Sigma}$ determined in (4.4), (4.10) and the definitions of distribution solutions and delta-shock solutions are immediately extended to these systems.

Indeed, these systems satisfy (4.1), but not (4.17); while u is a function of w, it is not a function of $w/(a\cdot w)$. Thus (4.20) does not apply.

For these systems, incomplete regularization provides approximate solutions assuming values everywhere within D, and satisfying the entropy conditions (3.11), (3.19), with only mild conditions on f, g, and the initial data. Explicitly,

$$\begin{aligned} &\rho_t^\varepsilon + (\rho^\varepsilon u^\varepsilon)_x = 0 \\ &g(u^\varepsilon)_t + (f(u^\varepsilon))_x = \varepsilon g(u^\varepsilon)_{xx},\ \varepsilon > 0 \end{aligned} \tag{4.31}$$

result in u^ε uniformly bounded and of bounded variation by a maximum principle argument. As $\varepsilon \downarrow 0, u^\varepsilon(\cdot, t) \to u^0(\cdot, t)$ boundedly in measure on $\mathbb{R}$, pointwise in t, and u^0 is an ordinary weak solution of the second equation in (4.28). Thus $M_i^{(2)}$ vanishes identically in (3.20), and the second component of (3.59) reduces to the ordinary Rankine-Hugoniot relation

$$\begin{aligned} &x_{i,t}(t)(g(u^0(x_i(t)+0,t)) - g(u^0(x_i(t)-0,t)) \\ &\quad = f(u^0(x_i(t)+0,t)) - f(u^0(x_i(t)-0,t)). \end{aligned} \tag{4.32}$$

With the speed and location of the discontinuities thus determined, the first component of $dM_i(t)dt$ in (3.59) in general does not vanish, so the ordinary shocks in the weak solution of the second equation in (4.28) become delta-shocks for (4.28) regarded as a pair of equations.

Interest in such systems (4.28) may be limited, because of the obvious reduction to scalar equations [BG]. Nevertheless, the treatment of such systems motivates and closely parallels the solution of some initial value problems for systems admitting singular shocks, described in chapter 3.

The example of zero pressure gas dynamics (3.1), however, illustrates the inherent limitations of this approach. An incompletely regularized form of (3.1) of the form

$$\rho_t^\varepsilon + (\rho^\varepsilon u^\varepsilon)_x = 0 \tag{4.33}$$

$$(\rho^\varepsilon u^\varepsilon)_t + (\rho^\varepsilon (u^\varepsilon)^2)_x = \varepsilon u_{xx}^\varepsilon,\ \varepsilon > 0, \tag{4.34}$$

has been used to obtain a viscous structure for the corresponding delta-shocks [TZZ, YL]. However, if (4.34) is replaced by

$$(\rho^\varepsilon u^\varepsilon)_t + (\rho^\varepsilon (u^\varepsilon)^2)_x = \varepsilon \rho^\varepsilon u_{xx}^\varepsilon, \tag{4.35}$$

the corresponding solutions are changed substantially. Indeed, solutions of (4.33), (4.35) are equivalent to those of (4.33) and

$$u_t^\varepsilon + \frac{1}{2}(u^\varepsilon)_x^2 = \varepsilon u_{xx}^\varepsilon, \tag{4.36}$$

and the differences between the limiting solutions of (4.33), (4.36) and the delta-shocks solutions of (3.1), for the same initial data, are substantial [BG, S3].

5. Entropy inequalities

Given a convex entropy density U with corresponding entropy flux F and a sequence w^ε obtained, for example by (possibly incomplete) regularization of (1.1), we have

$$U(w^\varepsilon)_t + F(w^\varepsilon)_x \le 0 \tag{5.1}$$

in the sense of distributions. In this generality one cannot pass to the limit as $\varepsilon \downarrow 0$; with w^0 given by (3.20), $U(w^0), F(w^0)$ are generally not defined.

THEOREM 5.1. *Assume that (5.1) holds for each* $\varepsilon > 0$, *that* U *is nonnegative and that the assumptions of theorem 3.3 hold, with* $w^\varepsilon \rightharpoonup w^0$ *as* $\varepsilon \downarrow 0, w^0$ *given in (3.20). Then*

$$U(w^\varepsilon(\cdot,t)) \rightharpoonup U(\tilde{w}(\cdot,t)) + \sum_i U_i(t)\chi_{I_i}(t)\delta(x - x_i(t)) \tag{5.2}$$

weakly in the space of measures on $\mathbb{R}$, *pointwise in* t. *The "singular entropy" functions* $U_i : I_i \to \mathbb{R}$ *satisfy*

$$\begin{aligned} \frac{dU_i(t)}{dt} &\leq x_{i,t}(t)\Big[U(\tilde{w}(x_i(t)+0,t)) - U(\tilde{w}(x_i(t)-0,t))\Big] \\ &\quad - F(\tilde{w}(x_i(t)+0,t)) + F(\tilde{w}(x_i(t)-0,t)). \end{aligned} \tag{5.3}$$

weakly in I_i.

At points $(\hat{x},\hat{t})$ *where singularities collide*

$$\sum_{j:x_j(\hat{t}+0)=\hat{x}} U_j(\hat{t}+0) \leq \sum_{i:x_i(\hat{t}-0)=\hat{x}} U_i(\hat{t}-0). \tag{5.4}$$

PROOF. The proof is very similar to that of theorem 3.3, with (5.1) replacing (2.1), and so will only be sketched. In compact regions of $\mathbb{R} \times \mathbb{R}_+$ not containing singularities

$$U(w^\varepsilon) \to U(\tilde{w}), F(w^\varepsilon) \to F(\tilde{w}) \tag{5.5}$$

boundedly in measure as $\varepsilon \downarrow 0$, and we recover the usual inequality

$$U(\tilde{w})_t + F(\tilde{w})_x \leq 0 \tag{5.6}$$

weakly, from (5.1).

Using U nonnegative and (5.1), as in the proof of theorem 3.1, the $U(u^\varepsilon(\cdot,t))$ are locally bounded measures on $\mathbb{R}$, pointwise in t, uniformly with respect to ε. Taking a subsequence as necessary at each t, (5.2) follows at least for these subsequences. Now considering the moments of (5.1) with a sequence of test functions θ such that each θ_x vanishes in an open neighborhood of $\{(x_i(t),t), t \in I_i\}$, we recover (5.3). Such an inequality (5.3) could not hold if the function $U_i(t)$ obtained in (5.2) were not unique, so the limit (5.2) is unique and no subsequences were in fact necessary. Finally the proof of (5.4) is analogous to that of (3.60). □

The values assumed by the right side of (5.3) are significantly different from those obtained for ordinary weak solutions of (1.1). For $w_- \in D$ fixed, $s \in \mathbb{R}, w \in D$, we denote the corresponding "Rankine-Hugoniot deficit" by e,

$$e(s,w,w_-) \stackrel{\text{def}}{=} s(w - w_-) - q(w) + q(w_-). \tag{5.7}$$

Denoting differentiation along an arbitrary curve in D by dots, we have from (5.7)

$$\dot{e} = (s - q_w)\dot{w} + \dot{s}(w - w_-). \tag{5.8}$$

Taking the inner product with the U_w, recalling $U_w q_w = F_w$,

$$\begin{aligned} U_w \cdot \dot{e} &= s\dot{U} - \dot{F} + \dot{s}U_w(w - w_-) \\ &= [s(U - U_-) - (F - F_-)] + \dot{s}(U_w \cdot (w - w_-) - U + U_-) \end{aligned} \tag{5.9}$$

abbreviating $U(w_-)$ by U_- etc. here and below.

Integrating (5.9) along a path connecting w_- with some fixed $w_+ \in D$, we obtain the corresponding "entropy drop" as the sum of two Riemann-Stieltjes integrals

$$E(w_-, w_+, s) \stackrel{\text{def}}{=} -s(U_+ - U_-) + F_+ - F_-$$

$$(5.10) \qquad = \int_{w_-}^{w_+} [U_w(w) \cdot (w - w_-) - U(w) + U_-]ds - U_w(w) \cdot de.$$

For U convex, the expression in brackets in (5.10) is nonnegative. For $w_+ \in \Gamma(w_-)$, one takes a path within $\Gamma(w_-)$ connecting w_- with w_+, along which e vanishes, and the familiar expression is recovered.

For the model problem for singular shocks (3.2), the two terms in (5.10) are of opposite sign, and the sign of $E(w_-, w_+, s)$ is in general not known a priori.

The singular entropy U_i behaves quite differently for delta-shocks than for singular shocks.

THEOREM 5.2. *Given a system (1.1) satisfying (3.18) and (4.1), assume* $(v^\varepsilon, \hat{w}^\varepsilon, \hat{q}^\varepsilon) : \mathbb{R} \times \mathbb{R}_+ \to \Sigma$ *converges boundedly in measure as* $\varepsilon \downarrow 0$ *to* $(v^0, \hat{w}^0, \hat{q}^0) : \mathbb{R} \times \mathbb{R}_+ \to \bar{\Sigma}$, *a single-valued delta-shock solution of (1.1). Assume that there exists a function* $\bar{U} : D \to \mathbb{R}$ *such that for any fixed* $\hat{w} \in D$,

$$(5.11) \qquad \lim_{\delta \downarrow 0} \delta U(\frac{\hat{w}}{\delta}) = \bar{U}(\hat{w}).$$

Then on each singularity

$$(5.12) \qquad U_i(t) = a \cdot M_i(t) \bar{U}(\frac{M_i(t)}{a \cdot M_i(t)}).$$

PROOF. As $\varepsilon \downarrow 0, U(w^\varepsilon)$ converges weakly as shown in (5.2), and with $x^\varepsilon(y,t)$ determined from v^ε as in (4.6), for each $t, \theta \in C_0(\mathbb{R} \to \mathbb{R})$

$$(5.13) \qquad \int_{\mathbb{R}} U(w^\varepsilon(x,t))\theta(x)dx = \int_{\mathbb{R}} U\Big(\frac{\hat{w}^\varepsilon(y,t)}{v^\varepsilon(y,t)}\Big)\theta(x^\varepsilon(y,t))v^\varepsilon(y,t)dy$$

using (4.7). Now for $y_-(t) < y_+(t)$ as determined in (3.49), as $\varepsilon \downarrow 0$,

$$(5.14) \qquad x^\varepsilon(y,t) \to x_i(t)$$

$$(5.15) \qquad v^\varepsilon(y,t) \downarrow 0$$

and

$$(5.16) \qquad \hat{w}^\varepsilon(y,t) \to \frac{M_i(t)}{a \cdot M_i(t)}$$

using the assumption that the solution is single valued, (3.50) and (4.13). Finally using the assumption (5.11), that the leading part of $U(w)$ is homogeneous of degree one in $|w|$, from (5.13)

$$(5.17) \qquad \lim_{\varepsilon \downarrow 0} \int_{y_-(t)}^{y_+(t)} v^\varepsilon(y,t) U\Big(\frac{\hat{w}^\varepsilon(y,t)}{v^\varepsilon(y,t)}\Big)\theta(x^\varepsilon(y,t))dy$$

$$= (y_+(t) - y_-(t))\bar{U}\Big(\frac{M_i(t)}{a \cdot M_i(t)}\Big)\theta(x_i(t))$$

and (5.12) follows from (5.2) and (5.17). □

For such a system (1.1) satisfying (4.17) additionally, we have (4.20) and (5.12) simultaneously, relating $x_{i,t}$, U_i, and M_i. Then (5.3) is interpreted as a dissipation of entropy; for example in the case of zero pressure gas dynamics (3.1), we have $a \cdot w = w_1, u = w_2/w_1, U = \frac{1}{2}w_2^2/w_1$ from (3.3), and thus $x_{i,t} = M_i^{(2)}/M_i^{(1)}, \bar{U} = (M_i^{(2)}/M_i^{(1)})^2/2$, and (5.3) is the well-known dissipation of kinetic energy.

6. Symmetries

Application of symmetries to the study of systems of conservation laws is well-known [BK]. A transformation of the form

$$w' = Zw + \eta_0, \quad Z \in M_{n\times n}(\mathbb{R}) \quad \text{nonsingular}, \ \eta_0 \in \mathbb{R}^n; \tag{6.1}$$

$$x' = \eta_3(\eta_1 x - \eta_2 t), \tag{6.2}$$

$$t' = \eta_3 t, \ \eta_1, \eta_2, \eta_3 \in \mathbb{R}, \ \eta_3 > 0, \ \eta_1 \neq 0 \tag{6.3}$$

implying

$$\frac{\partial}{\partial t'} = \frac{1}{\eta_3}\frac{\partial}{\partial t} + \frac{\eta_2}{\eta_1\eta_3}\frac{\partial}{\partial x}, \quad \frac{\partial}{\partial x'} = \frac{1}{\eta_1\eta_3}\frac{\partial}{\partial x}, \tag{6.4}$$

transforms weak solutions of (1.1) into weak solutions of

$$w'_{t'} + q(w')_{x'} = 0 \tag{6.5}$$

provided that

$$\begin{aligned} q(w') &= q(Zw + \eta_0) \\ &= Z(\eta_1 q(w) - \eta_2 w) + c, \end{aligned} \tag{6.6}$$

with $c \in \mathbb{R}^n$ independent of w.

For q satisfying (6.6), a sequence of approximate solutions satisfying (2.1) will also survive such a transformation. Distribution solutions of (1.1) are weak limits of such sequences. As the transformations (6.1), (6.6) are affine in $w, q(w)$, any such symmetry is expected to hold as well for solutions of the form (3.20), (3.57).

Often if not always, the symmetries form Lie groups, depending smoothly on a finite number of parameters. Euler systems are rich in this respect. For example, the system (1.16) admits a scaling symmetry, corresponding to

$$Z(\alpha) = \begin{pmatrix} 1/\alpha & 0 \\ 0 & 1 \end{pmatrix}, \eta_0 = 0, \eta_1(\alpha) = \alpha, \eta_2 = 0, \eta_3 = 1, \ \alpha > 0; \tag{6.7}$$

a reflection symmetry, corresponding to

$$Z = \begin{pmatrix} 1 & 0 \\ 0 & -1 \end{pmatrix}, \eta_0 = 0, \ \eta_1 = -1, \ \eta_2 = 0, \ \eta_3 = 1, \tag{6.8}$$

and a Galilean symmetry, corresponding to

$$Z(\beta) = \begin{pmatrix} 1 & 0 \\ -\beta & 1 \end{pmatrix}, \ \eta_0 = 0, \ \eta_1 = 1, \ \eta_2 = \beta, \ \eta_3 = 1, \ \beta \in \mathbb{R}. \tag{6.9}$$

The condition (6.6) suffices to obtain the invariance of distribution solutions of the form (3.20), (3.57), obtained as weak limits of approximations $w^\varepsilon, q(w^\varepsilon)$

satisfying (2.1-2.5). In particular, from (6.1-6.4), (6.6), we obtain

$$\begin{aligned} w^{\varepsilon'}_{t'} + q(w^{\varepsilon'})_{x'} &= \frac{1}{\eta_3} Z\big(w^{\varepsilon}_t + q(w^{\varepsilon})_x\big) \\ &\rightharpoonup 0 \text{ as } \varepsilon \downarrow 0. \end{aligned}$$

For a weak limit of the form (3.20),

$$\begin{aligned} w^{0'} &= \lim_{\varepsilon \downarrow 0} Z w^{\varepsilon} + \eta_0 \\ &= Z\tilde{w} + \eta_0 + \sum_i M_i'(t')\chi_{I_i'}(t')\delta\big(x' - x_i'(t')\big). \end{aligned} \tag{6.10}$$

From (6.2, 6.3) we may identify

$$I_i' = \eta_3 I_i$$

and

$$x_i'(t') = \eta_3(\eta_1 x(t) - \eta_2 t). \tag{6.11}$$

The corresponding singular mass can be obtained, for example, from (3.51)

$$M_i'(t') = \eta_1\eta_3 Z M_i(t). \tag{6.12}$$

Differentiating (6.11) with respect to t' and using (6.3), we obtain

$$x'_{i,t'} = \eta_1 x_{i,t} - \eta_2. \tag{6.13}$$

Identifying $x_{i,t}$ as the speed s associated with the Rankine-Hugoniot deficit given in (5.7), using (6.13), (6.1) and (6.6) we obtain

$$\begin{aligned} e'(s, w_+, w_-) &\stackrel{\text{def}}{=} e(s', w'_+, w'_-) \\ &= s'(w'_+ - w'_-) - q(w'_+) + q(w'_-) \\ &= Z(s(w_+ - w_-) - q(w_+) + q(w_-)) \\ &= Ze(s, w_+, w_-). \end{aligned} \tag{6.14}$$

In particular if $e(s, w_+, w_-) = 0$, meaning that w_+ and w_- can be connected by an (ordinary) shock of speed s, then w'_+ and w'_- can be connected by a shock of speed

$$s' = \eta_1 s - \eta_2. \tag{6.15}$$

Taking $|w_+ - w_-|$ small, we obtain the analogous relation for real characteristic speeds $\lambda_k(w), k = 1, \ldots, n$

$$\lambda_k(w') = \eta_1 \lambda_k(w) - \eta_2. \tag{6.16}$$

From (6.12) and (6.14), respectively, we deduce that if $\mathcal{M}$ is the set of all possible values of the singular mass and $\tilde{\mathcal{M}}$ is the set of all possible values of the Rankine-Hugoniot deficit associated with distribution solutions of (1.1), then the existence of such a symmetry implies

$$\eta_1\eta_3 Z\mathcal{M} = \mathcal{M} \tag{6.17}$$

and

$$Z\tilde{\mathcal{M}} = \tilde{\mathcal{M}}. \tag{6.18}$$

Systems such as the model problem for singular shocks (3.2) are identified with the exchange of conserved quantities in a system (1.1) [Ke]. Given an entropy density /flux U, F, for (1.1), such that

$$\frac{\partial U}{\partial w_n}(w) \neq 0 \quad \text{for all } w \in D, \tag{6.19}$$

smooth solutions of (1.1) correspond to those of

$$\tilde{w}_t + \tilde{q}(\tilde{w})_x = 0,\ \tilde{w} = \begin{pmatrix} w_1 \\ \vdots \\ w_{n-1} \\ U(w) \end{pmatrix}, \tilde{q}(\tilde{w}) = \begin{pmatrix} q_1(w) \\ \vdots \\ q_{n-1}(w) \\ F(w) \end{pmatrix}, \tag{6.20}$$

the condition (6.19) assuring that the map $w \to \tilde{w}$ is an isomorphism.

Whether a symmetry described by (6.1-6.3), (6.5) survives such an exchange depends on the entropy density U.

Given such a symmetry denote by

$$U'(w) = U(Zw + \eta_0), F'(w) = F(Zw + \eta_0), w \in D. \tag{6.21}$$

LEMMA 6.1. *The function U' is also an entropy density for the system (1.1), with corresponding entropy flux $(F' + \eta_2 U')/\eta_1$.*

PROOF. From $U_w(w)q_w(w) = F_w(w)$ we have

$$U_w(w')q_w(w') = F_w(w') \tag{6.22}$$

and as smooth solutions of (1.1) are smooth solutions of (6.5), multiplying (6.5) by $U_w(w')$ using (6.22) we have

$$U(w')_{t'} + F(w')_{x'} = 0. \tag{6.23}$$

Now using (6.4) and (6.21) in (6.23), we get

$$U'(w)_t + \frac{1}{\eta_1}(F'(w) + \eta_2 U'(w))_x = 0. \tag{6.24}$$

□

For a given symmetry, application of lemma 6.1 may not result in any "new" entropy functions. It may happen that for some $\kappa_0 \in \mathbb{R} \setminus \{0\}, \kappa \in \mathbb{R}^n,\ \underline{\kappa} \in \mathbb{R}$,

$$U'(w) = \kappa_0 U(w) + \kappa \cdot w + \underline{\kappa}. \tag{6.25}$$

LEMMA 6.2. *For $U'(w)$ of the form (6.25), and $s' = x'_{i,t'}$ given by (6.13),*

$$F'(w) = \eta_1(\kappa_0 F(w) + \kappa \cdot q(w)) - \eta_2(\kappa_0 U(w) + \kappa \cdot w) + c \tag{6.26}$$

and

$$E' = \eta_1(\kappa_0 E + \kappa \cdot e). \tag{6.27}$$

REMARK. In (6.27) e is the Rankine-Hugoniot deficit defined in (5.7) and E the entropy drop defined in (5.10), both for two arbitrary states $w_\pm \in D$ and $s \in \mathbb{R}$.

PROOF. From (6.21), differentiating with respect to w

$$U'_w(w) = U_w(w')Z,\ F'_w(w) = F_w(w')Z \tag{6.28}$$

and from (6.25)

$$U'_w(w) = \kappa_0 U_w(w) + \kappa \tag{6.29}$$

so

$$U_w(w') = (\kappa_0 U_w(w) + \kappa)Z^{-1}. \tag{6.30}$$

Differentiating (6.6) with respect to w

$$q_w(w')Z = Z(\eta_1 q_w(w) - \eta_2). \tag{6.31}$$

Now using (6.30) and (6.31) in (6.22),

$$\begin{aligned} F_w(w') &= (\kappa_0 U_w(w) + \kappa)(\eta_1 q_w(w) - \eta_2)Z^{-1} \\ &= (\kappa_0 \eta_1 F_w(w) + \eta_1 \kappa \cdot q_w(w) - \eta_2 \kappa_0 U_w(w) - \eta_2 \kappa)Z^{-1} \end{aligned} \tag{6.32}$$

and (6.26) follows easily from (6.32) and (6.28). Now

$$E' = s'(U'(w_+) - U'(w_-)) - F'(w_+) + F'(w_-);$$

with U' obtained from (6.25), F' from (6.26) and s' from (6.13), (6.27) follows. □

THEOREM 6.3. *Assume $U'(w)$ of the form (6.25) with*

$$\kappa_n = 0, \tag{6.33}$$

and in addition that

$$Z_{in} = 0, i = 1, \cdots, n-1. \tag{6.34}$$

Then corresponding to the symmetry (6.1-6.6) for the system (1.1) there is a symmetry described by $\hat{Z}, \hat{\eta}$ for the system (6.20) with

$$\begin{aligned} &\hat{Z}_{ij} = Z_{ij},\ i = 1, \ldots, n-1,\ j = 1, \ldots, n, \\ &\hat{Z}_{nj} = \kappa_j,\ j = 1, \ldots, n-1,\ \hat{Z}_{nn} = \kappa_0; \\ &\hat{\eta}_{0,j} = \eta_{0j},\ j = 1, \ldots, n-1,\ \hat{\eta}_{0n} = \underline{\kappa}; \\ &\hat{\eta}_1 = \eta_1, \hat{\eta}_2 = \eta_2, \hat{\eta}_3 = \eta_3. \end{aligned} \tag{6.35}$$

PROOF. It suffices to verify (6.6) for the system (6.20), i.e.

$$\tilde{q}(\tilde{w}') = \hat{Z}(\eta_1 \tilde{q}(\tilde{w}) - \eta_2 \tilde{w}) \tag{6.36}$$

with

$$\tilde{w}' = \hat{Z}\tilde{w} + \hat{\eta}_0. \tag{6.37}$$

From (6.34), the first $n-1$ components of (6.36) are identified with those of (6.6); it is here that (6.34) is needed. The last component of (6.36) is (6.26), using (6.35) for the $\hat{Z}_{nj}$ and (6.33). □

We observe that $w_n(\tilde{w})$ is also an entropy density satisfying (6.25) and (6.33) for the system (6.20); the corresponding entropy flux is $q_n(\tilde{w})$. The last component of (6.1) reads

$$w'_n = \sum_{j=1}^{n-1} Z_{nj} w_j + Z_{nn} w_n + \eta_{0n} \tag{6.38}$$

so (6.25) holds as well after such an interchange, with $U'(w)$ replaced by $w'_n(\tilde{w})$, $\kappa_0 = Z_{nn}, \kappa_j = Z_{nj}, j = 1, \ldots, n-1, \kappa_n = 0$.

For a given symmetry (6.1-6.6), the existence of an entropy density satisfying (6.25) is uncertain. For $\alpha \in \mathbb{R}$, suppose $Z(\alpha), \eta_0(\alpha), \eta_1(\alpha)$,

$\eta_2(\alpha), \eta_3(\alpha)$ determine a one-parameter Lie group of symmetries (satisfying (6.1-6.6)), parameterized so that

$$Z(0) = \text{identity}, \ |Z_\alpha(0)| + |\eta_{0,\alpha}(0)| > 0. \tag{6.39}$$

If (6.25) holds for each $\alpha \in \mathbb{R}$, i.e.

$$U(Z(\alpha)w + \eta_0(\alpha)) = \kappa_0(\alpha)U(w) + \kappa(\alpha) \cdot w + \underline{\kappa}(\alpha), \tag{6.40}$$

then differentiating (6.40) with respect to α and setting α equal to zero we find

$$U_w(w)(Z_\alpha(0)w + \eta_{0,\alpha}(0)) = \kappa_{0,\alpha}(0)U(w) + \kappa_\alpha(0) \cdot w + \kappa(\alpha) \tag{6.41}$$

for all $w \in D$. For some constants $\kappa_{0,\alpha}(0) \in \mathbb{R}, \kappa_\alpha(0) \in \mathbb{R}^n$, $\underline{\kappa}_\alpha(0) \in \mathbb{R}$, (6.41) is a necessary condition on U. Applying the criterion (6.41) to the system (1.16), for example, we find that there is no entropy of the form (6.25) for the symmetries determined by (6.7), but that the entropy (1.25) satisfies (6.41) for the symmetries determined by (6.9), and indeed satisfies (6.25).

The model problem for singular shocks (3.2) is exceptional with regard to its symmetry group. In addition to a reflection symmetry (6.8), this system admits two one-parameter Lie groups

$$Z = \begin{pmatrix} 1 & 0 \\ 0 & 1 \end{pmatrix}, \ \eta_0 = \begin{pmatrix} 0 \\ \alpha \end{pmatrix}, \ \eta_1 = \eta_3 = 1, \eta_2 = 0 \tag{6.42}$$

and

$$Z = \begin{pmatrix} 1 & 0 \\ -\beta & 1 \end{pmatrix}, \eta_0 = \begin{pmatrix} -\beta \\ 0 \end{pmatrix}, \ \eta_1 = \eta_3 = 1, \eta_2 = \beta. \tag{6.43}$$

Up to a scaling, (3.2) is the unique strictly hyperbolic pair of conservation laws admitting the two symmetry groups (6.42), (6.43) [Kl, S6].

These symmetries are "inherited" from the system (3.7), by application of theorem 6.3. The system (3.7) admits symmetries (6.8), (6.43) and the obvious scaling of ρ,

$$Z = \begin{pmatrix} \alpha & 0 \\ 0 & \alpha \end{pmatrix}, \ \eta_0 = 0, \eta_1 = \eta_3 = 1, \ \eta_2 = 0. \tag{6.44}$$

For all $\alpha, \beta \in \mathbb{R}$, the matrices Z in (6.42), (6.43) commute, as do those in (6.43), (6.44). This will be used in the discussion of the necessity of singular shocks and of the solution of the associated Cauchy problem in chapter 3.

The symmetry group of a given system (1.1) sometimes assists in the determination of the set $\mathcal{M}$ of possible values of singular mass. All systems of the form (1.1) admit trivial symmetries of the form $Z = I_{n\times n}, \eta_0 = 0, \eta_1 = 1, \eta_2 = 0, \eta_3 = \alpha, \alpha > 0$. The corresponding transformation $w' = w, x' = \alpha x, t' = \alpha t$ implies $\mathcal{M}' = \alpha\mathcal{M}$ from (6.17). Preservation of these symmetries thus requires $\mathcal{M}$ a cone. For $\mathcal{M}$ a cone with $\dim \mathcal{M} < n$ and Z not a scalar multiple of the identity matrix, $\mathcal{M}$ is contained in a proper invariant subspace of Z. These conditions are typical in the case of singular shocks; for example, applications to the system (3.2) with Z obtained from (6.43) shows that for this system $\mathcal{M}$ is a subset of the v-axis.

The analysis of distribution solutions of a given system (1.1) requires identification of the leading terms in the flux-function $q(w)$ for large $|w|$, and of the rate of growth of $|q(w)|$ as $|w|$ increases. This is accomplished by considering scalings of w satisfying (6.6) in the limit as $|w| \to \infty$.

DEFINITION. **The system (1.1) admits a symmetry at infinity if there exists $\tilde{Z} \in C((0,1) \to M_{n\times n}(\mathbb{R})$ nonsingular), $\tilde{\eta}_1 \in C((0,1] \to [1,\infty))$ with the following properties.**

$$\tilde{Z}(1) = I_{n\times n}, \tilde{\eta}_1(1) = 1, |\tilde{Z}(\varepsilon)| \to \infty \quad \textbf{as } \varepsilon \downarrow 0 \tag{6.45}$$

$$\tilde{Z}(\varepsilon_1\varepsilon_2) = \tilde{Z}(\varepsilon_1)\tilde{Z}(\varepsilon_2), \tilde{\eta}_1(\varepsilon_1\varepsilon_2) = \tilde{\eta}_1(\varepsilon_1)\tilde{\eta}_1(\varepsilon_2) \quad \textbf{for all } \varepsilon_1, \varepsilon_2 \in (0,1] \tag{6.46}$$

$$\tilde{Z}(\varepsilon)w \in D, \quad \textbf{for all } w \in D, \varepsilon \in (0,1] \tag{6.47}$$

$$q^L(w) \stackrel{\textbf{def}}{=} \lim_{\varepsilon\downarrow 0} \frac{1}{\tilde{\eta}_1(\varepsilon)} \tilde{Z}^{-1}(\varepsilon) q(\tilde{Z}(\varepsilon)w) \tag{6.48}$$

exists for all $w \in D$.

From (6.46) and (6.48), the "leading term" q^L satisfies a symmetry of the form

$$q^L(\tilde{Z}(\varepsilon)w) = \tilde{\eta}_1(\varepsilon)\tilde{Z}(\varepsilon)q^L(w). \tag{6.49}$$

All of the known systems (1.1) admitting distribution solutions admit such a symmetry at infinity. In the case of systems admitting single-valued delta-shock solutions, with (4.17) holding, one has necessarily

$$\tilde{\eta}_1(\varepsilon) = 1, \tag{6.50}$$

and

$$\tilde{Z}(\varepsilon) = I_{n\times n}/\varepsilon. \tag{6.51}$$

For such systems q^L is just the term in q homogeneous of degree one. We observe nonetheless that the system (4.28) satisfies (4.17) and for this system

$$\tilde{Z}(\varepsilon) = \begin{pmatrix} 1/\varepsilon & 0 \\ 0 & 1 \end{pmatrix}, \tilde{\eta}_1(\varepsilon) = 1. \tag{6.52}$$

For the model problem for singular shocks (3.2), [KK3, KK5, KK6]

$$\tilde{Z}(\varepsilon) = \begin{pmatrix} 1/\varepsilon & 0 \\ 0 & 1/\varepsilon^2 \end{pmatrix}, \tilde{\eta}_1(\varepsilon) = 1/\varepsilon \tag{6.53}$$

and

$$q^L\begin{pmatrix} u \\ v \end{pmatrix} = \begin{pmatrix} u^2 - v \\ u^3/3 \end{pmatrix}. \tag{6.54}$$

Although the existence of a symmetry at infinity is important in both the delta-shock case and the singular shock case, the application thereof is entirely different. In the case of delta-shocks, the symmetry at infinity is used to identify the speed of a delta-shock as a function of the singular mass. This corresponds to the function u for systems satisfying (4.17). In the case of singular shocks, in contrast, $\tilde{Z}$ is necessarily not a scalar multiple of the identity matrix. The singularity at infinity is used extensively in the analysis of the phase portraits of the corresponding "viscous profile system" in the region of large $|w|$.

7. Symmetry groups and phase space coordinates

Two methods have emerged for constructive solutions of the Cauchy problem for some systems (1.1), in the class of distribution solutions. Piecewise smooth initial data of bounded variation and not small oscillation is tacitly assumed. A shock-fitting algorithm is attractive in the case that all discontinuities are completely overcompressive. this often occurs for systems admitting delta-shock solutions, as described in chapter two.

Alternatively, a front tracking algorithm has been applied to find singular shock solutions for systems admitting exceptionally rich symmetry groups, most notably the model problem for singular shocks (3.2) [KLS]. In such cases, the symmetry group determines coordinates for phase space, with respect to which the system is translation invariant. Use of such coordinates greatly simplifies the analysis of the corresponding wave interaction problem.

The systems considered here admit an n-parameter Lie group of transformations $T_\alpha, \alpha \in \mathbb{R}^n$ of the form

(7.1) $$T_\alpha w = Z(\alpha)w + \eta_0(\alpha),\ Z(\alpha) \in M_{n\times n}(\mathbb{R}),\ \ \eta_0(x) \in \mathbb{R}^n;$$

(7.2) $$T_\alpha x = \eta_3(\alpha)(\eta_1(\alpha)x - \eta_2(\alpha)t),\ \eta_1(\alpha), \eta_3(\alpha) > 0,\ \ \eta_2(\alpha) \in \mathbb{R};$$

(7.3) $$T_\alpha t = \eta_3(\alpha)t.$$

These transformations are assumed to satisfy several properties. They form a commutative group under multiplication; for all $\alpha, \alpha' \in \mathbb{R}^n$ there exists $\beta \in \mathbb{R}^n$ such that

(7.4) $$T_\alpha T_{\alpha'} = T_{\alpha'} T_\alpha = T_\beta,$$

noting that in general $\beta \neq \alpha + \alpha'$.

These transformations are symmetries for the system (1.1), satisfying (6.6) for any $\alpha \in \mathbb{R}^n$,

(7.5) $$q\big(Z(\alpha)w + \eta_0(\alpha)\big) = Z(\alpha)\big(\eta_1(\alpha)q(w) - \eta_2(\alpha)w\big) + c(\alpha).$$

The functions $Z, \eta_i, i = 0, 1, 2, 3$ satisfy

(7.6) $$Z, \eta_i \in C^1(\mathbb{R}^n), Z(0) = I_{n\times n}, \eta_0(0) = 0, \eta_1(0) = \eta_3(0) = 1, \eta_2(0) = 0.$$

For any fixed $w \in D$, the mapping

(7.7) $$\alpha \to T_\alpha w$$

is an isomorphism of $\mathbb{R}^n$ onto D. Finally, the "infinitesimal matrix"

(7.8) $$Q(w) = \frac{\partial T_\alpha w}{\partial \alpha}\bigg|_{\alpha=0}$$

is nonsingular for all $w \in D$.

Examples include the model problem for singular shocks (3.2) with $D = \mathbb{R}^2, w = \begin{pmatrix} u \\ v \end{pmatrix}, \alpha = \begin{pmatrix} \alpha_1 \\ \alpha_2 \end{pmatrix}$, and

(7.9) $$Z(\alpha) = \begin{pmatrix} 1 & 0 \\ -\alpha_2 & 1 \end{pmatrix}, \eta_0(\alpha) = \begin{pmatrix} -\alpha_2 \\ \alpha_1 \end{pmatrix}, \eta_1 = \eta_3 = 1, \eta_2(\alpha) = \alpha_2, Q(w) = \begin{pmatrix} 0 & -1 \\ 1 & -u \end{pmatrix},$$

is "isothermal gas dynamics" (3.7) or "zero pressure gas dynamics" (3.1), for either of which $D = \{\rho > 0\}$, $w = \begin{pmatrix} \rho \\ \rho u \end{pmatrix}$, and

$$Z(\alpha) = \begin{pmatrix} e^{\alpha_1} & 0 \\ -\alpha_2 e^{\alpha_1} & e^{\alpha_1} \end{pmatrix}, \eta_0 = 0, \eta_1 = \eta_3 = 1,\ \eta_2(\alpha) = \alpha_2, Q(w) = \begin{pmatrix} \rho & 0 \\ \rho u & -\rho \end{pmatrix}. \tag{7.10}$$

We remark that the conditions (7.4-7.8) do not uniquely determine α, even up to linear transformation. The example of zero pressure gas dynamics shows that such systems are not necessarily hyperbolic.

The nonuniqueness notwithstanding, the parameters α are attractive coordinates for phase space.

THEOREM 7.1. *Assume (7.1-7.8) hold. Then there exists a smooth isomorphism* γ *on* $\mathbb{R}^n$ *such that*

$$T_{\gamma(\alpha+\alpha')} = T_{\gamma(\alpha)} T_{\gamma(\alpha')} \tag{7.11}$$

for all $\alpha, \alpha' \in \mathbb{R}^n$.

REMARKS. Fix $w_0 \in D$ arbitrarily and determine $w(\alpha)$ by

$$w(\alpha) = T_{\gamma(\alpha)} w_0 \tag{7.12}$$

then from (7.11) and (7.12)

$$w(\alpha + \beta) = T_{\gamma(\beta)} w(\alpha) \tag{7.13}$$

so using the α as coordinates for phase space, translations correspond to symmetries for the system (1.1).

In general $\gamma(\alpha) \neq \alpha$: for example for the system (3.2), one may compute $\gamma_1 = \alpha_1 + \alpha_2^2/2, \gamma_2 = \alpha_2$. However it is true that for $|\alpha|$ small

$$|\gamma(\alpha) - \alpha| = o(|\alpha|). \tag{7.14}$$

Up to a linear transformation, γ is unique. This will emerge in the proof of theorem 7.4 below.

PROOF. For m a positive integer, $w \in D, \alpha \in \mathbb{R}^n$ denote by

$$\psi_{\alpha,m} w = (T_{\alpha/m})^m w. \tag{7.15}$$

From (7.6)

$$|T_{\alpha/m} w| = (1 + O(\frac{|\alpha|}{m}))w, \tag{7.16}$$

so that for α, w fixed, the $|\psi_{\alpha,m} w|$ are bounded uniformly in m.

From the group property (7.4), there is an α_m independent of w such that

$$\psi_{\alpha,m} w = T_{\alpha_m} w. \tag{7.17}$$

There exists a subsequence $\{m_j\}_{j=1}^{\infty}$, possibly depending on α and on w, such that the sequence $\psi_{\alpha,m_j} w$ converges as $j \to \infty$. Using (7.7) and (7.17), this can only happen if the sequence $\{\alpha_{m_j}\}$ converges. As the α_m are independent of w, so is the limit

$$\gamma(\alpha) \stackrel{\text{def}}{=} \lim_{j \to \infty} \alpha_{m_j}. \tag{7.18}$$

Next we establish convergence of the entire sequence $\{\alpha_m\}_{m=1}^{\infty}$ by establishing uniqueness of the limit of any convergent subsequence $\{\alpha_{m_j}\}$. Using (7.8) and (7.6), we have

$$T_{\alpha/m}w = w + Q(w)\frac{\alpha}{m} + o\left(\frac{|\alpha|}{m}\right). \tag{7.19}$$

Using (7.15), (7.17), (7.19), for any convergent subsequence $\{\alpha_{m_j}\}$ we may identify the limit

$$\lim_{j\to\infty} \psi_{\alpha,m_j} w = \omega(1) \tag{7.20}$$

where $\omega(\xi), \xi \in \mathbb{R}$ is obtained from a differential equation

$$\omega_\xi = Q(\omega)\alpha, \xi \in \mathbb{R}; \omega(0) = w. \tag{7.21}$$

From (7.8)

$$|Q(\omega)| \le c(1 + |\omega|), \tag{7.22}$$

so the solution of (7.21) exists and is unique for all $\xi \in \mathbb{R}$.

Using (7.7), uniqueness of the limits (7.20) for any convergent subsequence $\{\alpha_{m_j}\}$ implies uniqueness of the limits of the subsequences $\{\alpha_{m_j}\}$, so (7.18) can be improved to

$$\gamma(\alpha) = \lim_{m\to\infty} \alpha_m, \tag{7.23}$$

equivalently

$$T_{\gamma(\alpha)} = \lim_{m\to\infty} (T_{\alpha/m})^m, \tag{7.24}$$

using (7.15) and (7.17).

From (7.19), for any $\alpha, \beta \in \mathbb{R}^n$

$$\begin{aligned}
T_{\alpha/m}T_{\beta/m}w &= T_{\beta/m}w + Q(T_{\beta/m}w)\frac{\alpha}{m} + o(1/m) \\
&= w_+Q(w)\frac{\beta}{m} + Q(w + Q(w)\frac{\beta}{m} + o(1/m))\frac{\alpha}{m} + o(1/m) \\
&= w + Q(w)\left(\frac{\alpha+\beta}{m}\right) + o(1/m) \\
&= T_{(\alpha+\beta)/m}w + o(1/m) \qquad (7.25)
\end{aligned}$$

so that

$$(T_{\alpha/m}T_{\beta/m})^m w = (T_{(\alpha+\beta)/m})^m w + o(1). \tag{7.26}$$

Now using (7.24), (7.4) and (7.26)

$$\begin{aligned}
T_{\gamma(\alpha)}T_{\gamma(\beta)}w &= \lim_{m\to\infty} (T_{\alpha/m})^m \lim_{k\to\infty} (T_{\beta/k})^k w \\
&= \lim_{m\to\infty} (T_{\alpha/m})^m (T_{\beta/m})^m w \\
&= \lim_{m\to\infty} (T_{\alpha/m}T_{\beta/m})^m w \\
&= \lim_{m\to\infty} \left[(T_{(\alpha+\beta)/m})^m w + o(1)\right] \\
&= T_{\gamma(\alpha+\beta)}w \qquad (7.27)
\end{aligned}$$

It remains to prove that the mapping $\alpha \to \gamma(\alpha)$ is an isomorphism on $\mathbb{R}^n$. For any $\alpha \in \mathbb{R}^n, \gamma(\alpha)$ is obtained explicitly by solving (7.21) and identifying

$$\omega(1) = T_{\gamma(\alpha)}\omega(0), \tag{7.28}$$

recalling (7.15), (7.20) and (7.24). In view of (7.24), the obtained $\gamma(\alpha)$ is independent of the choice of $\omega(0)$. For any given $\gamma \in \mathbb{R}^n$, fix $w_0 \in \mathbb{R}^n$ and apply (7.12)

$$w(\alpha(\gamma)) = T_\gamma w_0. \tag{7.29}$$

From (7.7), the map $\gamma \to T_\gamma w_0$ is on isomorphism $\mathbb{R}^n$ onto D, so it will suffice to show that the map $\alpha \to w(\alpha)$ determined by (7.12) is an isomorphism of $\mathbb{R}^n$ onto D.

For small α, (7.21) becomes

$$\omega(1) = \omega(0) + Q(\omega(0))\alpha + o(|\alpha|), \tag{7.30}$$

whereas from (7.8), for small $|\gamma|$,

$$T_\gamma\omega(0) = \omega(0) + Q(\omega(0))\gamma + o(|\gamma|). \tag{7.31}$$

Comparing (7.30) with (7.31), using (7.28), proves (7.14), and in particular

$$\frac{\partial\gamma(\alpha)}{\partial\alpha}\Bigg|_{\alpha=0} = I_{n\times n}. \tag{7.32}$$

Having proved (7.11), it follows that (7.13) holds. Differentiating (7.13) with respect to β and setting $\beta = 0$, using (7.8) and (7.32) it follows that the map $\alpha \to w(\alpha)$ determined by (7.12) satisfies

$$w_\alpha(\alpha) = Q(w(\alpha)). \tag{7.33}$$

From the nonsingularity of Q, there is a locally defined inverse function $\alpha(w)$ with

$$\alpha_w(w) = Q(w)^{-1}. \tag{7.34}$$

From (7.6) and (7.7), D is necessarily simply connected, so the integral

$$\alpha(w) = \int_{w_0}^{w} Q(z)^{-1}dz \tag{7.35}$$

is defined, i.e. independent of path, for any $w \in D$. Since the inverse function $\alpha(w)$ is explicitly given in (7.35), it follows that (7.12) is indeed an isomorphism of $\mathbb{R}^n$ onto D. □

For the remainder of this section, we assume a system (1.1) endowed with a symmetry group satisfying (7.1-7.8). We obtain results on the structure of such systems and on the functional form of the symmetry group. We also consider the survival of such a symmetry group under an exchange of conserved quantities.

LEMMA 7.2. *In the coordinates α, the eigenvector fields corresponding to real distinct characteristic speeds are straight lines.*

REMARKS. If $\lambda_k(w_0)$ is real and distinct for some $k = 1, \ldots, n$ and any $w_0 \in D$, then from (6.16) and (7.7), $\lambda_k(w)$ is real and distinct for all $w \in D$. If $\underline{r}_k(w)$ is the corresponding eigenvector of $q_w(w)$, then from (7.33) in the coordinates α the corresponding eigenvector is

(7.36) $$r_k(\alpha(w)) = Q(w)^{-1}\underline{r}_k(w)..$$

PROOF. For any $w_0 \in D$, regardless of the behavior of the k-th characteristic family in a neighborhood of w_0, there exists $\alpha(x,t)$ assuming values in a neighborhood of $\alpha(w_0)$ and corresponding to a smooth similarity solution of (1.1), ie. with α_x, α_t parallel to r_k at each point. Then for any fixed $\beta \in \mathbb{R}^n$, the function $\alpha(x,t) + \beta$ corresponds to a smooth similarity solution of the same form, assuming values in a neighborhood of $w(\alpha(w_0)+\beta) = T_\beta w_0$ using (7.13). Thus $r_k(\alpha(x,t)+\beta)$ must be parallel to

(7.37) $$(\alpha(x,t) + \beta)_x = \alpha_x(x,t)$$

which is parallel to $r_k(\alpha(x,t))$. So the direction of r_k is constant throughout $\mathbb{R}^n$. □

An immediate consequence is the following.

THEOREM 7.3. *Assume a system (1.1) strictly hyperbolic with dimension $n \geq 3$, and admitting a symmetry group satisfying (7.1-7.8). Then there exists a strict Riemann invariant for each characteristic family, and the α are linear combinations of strict Riemann invariants.*

REMARK. Such systems are thus rich in the sense of Serre [Se3].

PROOF. The vectors $r_k, k = 1, \ldots, n$, are linearly independent. Denote by R the nonsingular matrix obtained by assembling the r_k columnwise, and $b_j, j = 1, \ldots, n$, the j-th row of R^{-1}. Regarding b_j as a row vector and α as a column vector, $\mu_j(w) \stackrel{\text{def}}{=} b_j \cdot \alpha(w)$ is a strict Riemann invariant for characteristic family $j, j = 1, \cdots, n$. And $\alpha = R\mu$ where $\mu = R^{-1}\alpha$ is just the $\mu_j(w)$ assembled columnwise. □

Symmetry groups of the form (7.1-7.3) satisfying (7.4-7.8) are invariably found by combining n "independent" one-parameter symmetry groups, each obtained by inspection for a given system (1.1). For example, for the system (3.2), the symmetry group determined by (7.9) is obviously obtained by combining (6.42), (6.43). Assume, then, that for $j = 1, \ldots, n$ and any α_j, we have $Z(\alpha_j)$ nonsingular, $\eta_1(\alpha_j) > 0, \eta_3(\alpha_j) > 0, \eta_2(\alpha_j) \in \mathbb{R}, \eta_0, (\alpha_j) \in \mathbb{R}^n$ such that (6.6) holds. Assume in addition that for any $\alpha_j, \alpha_j' \in \mathbb{R}$, there exists $\beta_j(\alpha_j, \alpha_j') \in \mathbb{R}$ such that

(7.38) $$T_{\alpha_j} T_{\alpha_j'} = T_{\alpha_j'} T_{\alpha_j} = T_{\beta_j}.$$

We seek an n-parameter symmetry group in the obvious manner, identifying

(7.39) $$\alpha = \begin{pmatrix} \alpha_1 \\ \vdots \\ \alpha_n \end{pmatrix}, T_\alpha = \prod_{j=1}^{n} T_{\alpha_j}.$$

A necessary condition to obtain a commutative symmetry group is that the T_α thus obtained commute, i.e. the first equality in (7.4) holds. From

(7.40) $$T_\alpha T_{\alpha'} w = T_{\alpha'} T_\alpha w$$

for any $w \in D$, using (7.1) one readily obtains

$$Z(\alpha)Z(\alpha') = Z(\alpha')Z(\alpha), \tag{7.41}$$

and

$$Z(\alpha)\eta_0(\alpha') + \eta_0(\alpha) = Z(\alpha')\eta_0(\alpha) + \eta_0(\alpha'); \tag{7.42}$$

requiring

$$T_\alpha T_{\alpha'} x = T_{\alpha'} T_\alpha x \tag{7.43}$$

similarly gives

$$\eta_1(\alpha)\eta_2(\alpha') + \eta_2(\alpha) = \eta_1(\alpha')\eta_2(\alpha) + \eta_2(\alpha') \tag{7.44}$$

as necessary conditions.

As expected, the existence of a multiplicative group, i.e. the second equality in (7.4), follows from (7.38), the commutivity of the T_α, and (7.39)

$$\begin{aligned} T_\alpha T_{\alpha'} &= \left(\prod_{j=1}^{n} T_{\alpha_j}\right)\left(\prod_{k=1}^{n} T_{\alpha'_k}\right) \\ &= \prod_{j=1}^{n} (T_{\alpha_j} T_{\alpha'_j}) \\ &= \prod_{j=1}^{n} T_{\beta_j} \\ &= T_\beta. \end{aligned} \tag{7.45}$$

Using (7.45), for any $\alpha, \alpha' \in \mathbb{R}^n$ we find $Z(\beta), \eta_i(\beta), i = 0, 1, 2, 3$ explicitly. In particular, $Z(\beta), \eta_0(\beta), \eta_2(\beta)$ are identified with either side of (7.41), (7.42), (7.44), respectively, while (7.43) and (7.45) imply

$$\eta_1(\beta) = \eta_1(\alpha)\eta_1(\alpha') \tag{7.46}$$

and $T_\alpha T_{\alpha'} t = T_\beta t$ implies

$$\eta_3(\beta) = \eta_3(\alpha)\eta_3(\alpha'). \tag{7.47}$$

Since each of the T_{α_j} satisfies (6.6), T_α obtained from (7.39) will satisfy (6.5). In any given specific example, the remaining assumptions of theorem 7.1, namely (7.6-7.8), are easily checked by inspection. In particular, the meaning of "independent" one-parameter symmetry groups above is that (7.7) and (7.8) hold simultaneously.

A representation theorem holds for such n-parameter symmetry groups.

THEOREM 7.4. *Assume a symmetry group of the form (7.1-7.3) satisfying (7.4-7.8). Let $\gamma(\alpha)$ be the function as determined by Theorem 7.1. Then there exist commuting $n \times n$ matrices $Z_1, \ldots, Z_n$, parallel n-vectors a_1, a_2, and an n-vector a_3*

such that

$$Z(\gamma(\alpha)) = \exp\left(\sum_{j=1}^{n} \alpha_j Z_j\right) \tag{7.48}$$

$$\eta_1(\gamma(\alpha)) = \exp(a_1 \cdot \alpha) \tag{7.49}$$

$$\eta_2(\gamma(\alpha)) = \begin{cases} a_2 \cdot \alpha, a_1 = 0 \\ \dfrac{a_1 \cdot a_2}{|a_1|^2}(\exp(a_1 \cdot \alpha) - 1),\ a_1 \neq 0 \end{cases} \tag{7.50}$$

$$\eta_3(\gamma(\alpha)) = \exp(a_3 \cdot \alpha). \tag{7.51}$$

REMARKS. In (7.49-7.51) and below, dots denote the real inner production on $\mathbb{R}^n$. Parallel vectors a_1, a_2 mean that one is a scalar multiple of the other - either or both may be zero. Various conditions on $\tilde{\eta}_0(\gamma(\alpha))$ are obtained from (7.57) below and the assumption (7.8).

PROOF. Denote by $\tilde{T}_\alpha = T_{\gamma(\alpha)}, \tilde{Z}(\alpha) = Z(\gamma(\alpha)), \tilde{\eta}_i(\alpha) = \eta_i(\gamma(\alpha)), i = 0, 1, 2, 3.$ In this notation, using (7.11), (7.4) becomes

$$\tilde{T}_{\alpha'}\tilde{T}_\alpha = \tilde{T}_\alpha\tilde{T}_{\alpha'} = \tilde{T}_{\alpha+\alpha'} \tag{7.52}$$

and (7.1-7.3) are

$$\tilde{T}_\alpha w = \tilde{Z}(\alpha)w + \tilde{\eta}_0(\alpha) \tag{7.53}$$

$$\tilde{T}_\alpha x = \tilde{\eta}_3(\alpha)(\tilde{\eta}_1(\alpha)x - \tilde{\eta}_2(\alpha)t) \tag{7.54}$$

$$\tilde{T}_\alpha t = \tilde{\eta}_3(x)t. \tag{7.55}$$

Combining (7.52) and (7.53) we easily obtain

$$\tilde{Z}(\alpha + \alpha') = \tilde{Z}(\alpha)\tilde{Z}(\alpha') = \tilde{Z}(\alpha')\tilde{Z}(\alpha) \tag{7.56}$$

and

$$\begin{aligned} \tilde{\eta}_0(\alpha + \alpha') &= \tilde{Z}(\alpha)\tilde{\eta}_0(\alpha') + \tilde{\eta}_0(\alpha) \\ &= \tilde{Z}(\alpha')\tilde{\eta}_0(\alpha) + \tilde{\eta}_0(\alpha'). \end{aligned} \tag{7.57}$$

As the $\tilde{Z}$ are nonsingular, the logarithm of (7.56) gives (7.48). As the $\tilde{Z}$ all commute, so do the $Z_i, i = 1, \ldots, n$.

Combining (7.52) and (7.55) gives

$$\tilde{\eta}_3(\alpha + \alpha') = \tilde{\eta}_3(\alpha)\tilde{\eta}_3(\alpha'), \tag{7.58}$$

from which (7.51) follows.

Combining (7.52) and (7.54), using (7.58) we similarly obtain

$$\tilde{\eta}_1(\alpha + \alpha') = \tilde{\eta}_1(\alpha)\tilde{\eta}_1(\alpha'), \tag{7.59}$$

from which (7.49) follows, and

$$\tilde{\eta}_2(\alpha + \alpha') = \tilde{\eta}_1(\alpha)\tilde{\eta}_2(\alpha') + \tilde{\eta}_2(\alpha). \tag{7.60}$$

The limit as $\alpha' \to 0$ in (7.60) gives

$$\tilde{\eta}_{2,\alpha} = \tilde{\eta}_1(\alpha)\tilde{\eta}_{2,\alpha}(0); \tag{7.61}$$

using (7.49), (7.61) is integrated to obtain (7.50). □

Reversing the proof, we obtain a converse statement of theorem 7.4; (7.48-7.51) and (7.57) together imply (7.52).

We note in addition that the transformations $\tilde{T}_\alpha$ satisfy (7.5-7.8), i.e. the assumptions of theorem 7.1. Using (7.32), the same matrix $Q(w)$ in (7.8) is obtained for $\tilde{T}_\alpha$ as for T_α. Thus the $\tilde{T}_\alpha$ may be viewed as reparameterization of the transformations T_α, in such a way that the function γ corresponding to $\tilde{T}_\alpha$ becomes the identity.

After such a reparameterization we shall have $Z(\alpha), \eta_1(\alpha), \eta_2(\alpha)$, $\eta_3(\alpha)$ of the explicit form shown in (7.48-7.51), $\eta_0(\alpha)$ obtained from (7.57), and $\gamma(\alpha) = \alpha$. The explicit form for T_α determines α, and thus γ, up to invertible linear transformation. Thus for any system (1.1) admitting a symmetry group satisfying the assumptions of theorem 7.1, the function $\gamma(\alpha)$ is independent of the specific parameterization of T_α, up to invertible linear transformation. Using (7.32), so is the matrix $Q(w)$ in (7.8), and thus from (7.35), so are the obtained phase space coordinates $\alpha(w)$.

Given a system (1.1), the existence of a symmetry group satisfying the assumptions of theorem 7.1 is obviously exceptional. However, given such a symmetry group, its existence often survives subsequent exchanges of the conserved quantities.

Assume a system (1.1) of dimension n equipped with an entropy density/flux $U(w), F(w)$, such that for all $w \in D$,

$$\frac{\partial U}{\partial w_n}(w) \neq 0. \tag{7.62}$$

In particular U does not have to be convex in w. Then smooth solutions of (1.1) assuming values in D correspond to smooth solutions of

$$w_t^X + q_x^X = 0 \tag{7.63}$$

assuming values in D^X, where

$$w^X(w) = \begin{pmatrix} w_1 \\ \vdots \\ w_{n-1} \\ U(w) \end{pmatrix}, q^X(w^X) = \begin{pmatrix} q_1(w) \\ \vdots \\ q_{n-1}(w) \\ F(w) \end{pmatrix}, D^X = \{w^X(w), w \in D\}. \tag{7.64}$$

THEOREM 7.5. *Assume a system (1.1) equipped with an entropy density/flux satisfying (7.62), and equipped with a symmetry group of the form (7.1-7.3), satisfying (7.4-7.8), and with*

$$Z_{in}(\alpha) = 0,\ i = 1,\ \ldots, n-1,\ \alpha \in \mathbb{R}^n. \tag{7.65}$$

Assume in addition that for all $w \in D, \alpha \in \mathbb{R}^n$

$$U(Z(\alpha)w + \eta_0(\alpha)) = \kappa_0(\alpha)U(w) + \kappa(\alpha) \cdot w + \underline{\kappa}(\alpha) \tag{7.66}$$

with $\kappa_0, \kappa, \underline{\kappa}$ of class C^1 satisfying

$$\kappa_0(0) = 1, \kappa(0) = 0, \underline{\kappa}(0) = 0 \tag{7.67}$$

and

$$\kappa_n(\alpha) = 0, \kappa_0(\alpha) > 0 \tag{7.68}$$

for all $\alpha \in \mathbb{R}^n$.

Then there exists a symmetry group satisfying the assumptions of theorem 7.1 for the system (7.63), given explicitly by

$$Z^X(\alpha) = \begin{pmatrix} & \vdots \\ Z_{ij}(\alpha) & \vdots\, 0 \\ i,j \leq n-1 & \vdots \\ \dots\dots\dots\dots\dots & \vdots\dots \\ \kappa_1(\alpha)\dots\kappa_{n-1}(\alpha) & \vdots\,\kappa_0(\alpha) \end{pmatrix} \qquad \eta_0^X(\alpha) = \begin{pmatrix} \eta_{0,1}(\alpha) \\ \vdots \\ \eta_{0,n-1}(\alpha) \\ \underline{\kappa}(a) \end{pmatrix}$$

$$\eta_i^X(\alpha) = \eta_i(\alpha), i = 1,2,3. \tag{7.69}$$

REMARKS. The assumption (7.65) in fact involves no loss of generality. A linear transformation on α will put the commuting $Z_1, \dots, Z_n$ in lower Jordan form simultaneously, thus assuring that (7.65) holds, using (7.48). The assumed form of $U(T_\alpha w)$ in (7.66) is motivated by lemma 6.1 and (6.25).

PROOF. Denote by

$$Xw = w^X(w), T_\alpha^X w^X = Z^X(\alpha)w^X + \eta_0^X(\alpha) \tag{7.70}$$

observing that X^{-1} is defined on D^X in view of (7.62).

From (7.69) and (7.70), it follows that for all $\alpha \in \mathbb{R}^n$, $w \in D$,

$$T_\alpha^X Xw = XT_\alpha w. \tag{7.71}$$

Thus for any $\alpha, \alpha' \in \mathbb{R}^n, \beta \in \mathbb{R}^n$ as obtained in (7.4) for the mapping T_α, we have

$$\begin{aligned} T_\alpha^X T_{\alpha'}^X w^X &= T_\alpha^X T_{\alpha'}^X XX^{-1}w^X \\ &= XT_\alpha T_{\alpha'} X^{-1}w^X \\ &= XT_\beta X^{-1}w^X \\ &= T_\beta^X XX^{-1}w^X \\ &= T_\beta^X w^X \end{aligned} \tag{7.72}$$

so (7.4) holds for the T_α^X with the same $\beta(\alpha, \alpha')$ as for T_α. By appeal to theorem 6.3 for each $\alpha \in \mathbb{R}^n$, using (7.65) and (7.68), it follows that (6.6) holds for T_α^X. (7.6) follows from (7.69) and (7.67).

For fixed w^X, the mapping of $\mathbb{R}^n$ into D^X determined by $\alpha \to T_\alpha^X w^X$ is easily identified as an isomorphism, as

$$\begin{aligned} T_\alpha^X w^X &= T_\alpha^X XX^{-1}w^X \\ &= XT_\alpha X^{-1}w^X \end{aligned} \tag{7.73}$$

and recalling that (7.7) holds for T_α by assumption.

Finally, to obtain (7.8) we again apply (7.71) to obtain, for arbitrary fixed $w^X \in D^X$,

$$
\begin{aligned}
Q^X(w^X) &\stackrel{\text{def}}{=} \frac{\partial}{\partial\alpha} T_\alpha^X w^X \Big|_{\alpha=0} \\
&= \frac{\partial}{\partial\alpha}(T_\alpha^X X X^{-1} w^X) \Big|_{\alpha=0} \\
&= \frac{\partial}{\partial\alpha}(X T_\alpha X^{-1} w^X) \Big|_{\alpha=0} \\
&= \frac{\partial X}{\partial w}(T_\alpha X^{-1} w^X) \frac{\partial}{\partial\alpha}(T_\alpha X^{-1} w^X) \Big|_{\alpha=0} \\
&= \frac{\partial X}{\partial w}(T_\alpha X^{-1} w^X) Q(X^{-1} w^X)
\end{aligned} \tag{7.74}
$$

Now $\partial X/\partial w$ is everywhere nonsingular from (7.70), (7.64) and (7.62), while Q is everywhere nonsingular and (7.8) holds for the T_α^X as well. □

From (7.35) and (7.74), we note that the same phase space coordinates $\alpha(w)$ are obtained for the original system (1.1) as $\alpha(w^X)$ obtained for the system (7.63). This is not surprising in the case where (1.1) (and thus (7.63)) are strictly hyperbolic, in view of theorem 7.3, as strict Riemann invariants survive such exchanges of conserved quantities.

We conclude this section with an example. The system

$$
\begin{aligned}
&\rho_t + (\rho u)_x = 0 \\
&u_t + (\frac{1}{2}u^2 + \log\rho)_x = 0, \quad D = \{\rho > 0\},
\end{aligned} \tag{7.75}
$$

is equipped with a (nonconvex) entropy

$$U(\rho, u) = \rho u, \quad F(\rho, u) = \rho u^2 + \rho \tag{7.76}$$

and a symmetry group of the form

$$Z(\alpha) = \begin{pmatrix} e^{\alpha_1} & 0 \\ 0 & 1 \end{pmatrix}, \eta_0(\alpha) = \begin{pmatrix} 0 \\ -\alpha_2 \end{pmatrix}, \ \eta_1 = \eta_3 = 1, \ \eta_2(\alpha) = \alpha_2. \tag{7.77}$$

Identifying w_n as u, we see that for U given in (7.76), we have (7.66) holding with

$$\kappa_0(\alpha) = e^{\alpha_1}, \ \ \kappa(\alpha) = (-\alpha_2 e^{\alpha_1} \ \ 0) \ \ \underline{\kappa}(\alpha) = 0 \tag{7.78}$$

thus satisfying (7.67) and (7.68). In this case the resulting system (7.63), (7.64), is isothermal gas dynamics (3.7), with symmetry group shown in (7.10).

Alternatively, the system (7.75) admits strictly convex entropy

$$U = \frac{1}{2}u^2 - \log\rho, \quad F = uU. \tag{7.79}$$

We cannot exchange U given by (7.79) with u, as (7.62) would fail, so we reverse the order of the equations in (7.75) and identify w_n as ρ. Then (7.62) holds and we readily verify (7.66) holding with

$$\kappa_0(\alpha) = 1, \ \kappa(\alpha)(-\alpha_2 \ 0), \underline{\kappa}(\alpha) = \frac{1}{2}\alpha_2^2 - \alpha_1. \tag{7.80}$$

In this case the system (7.63), (7.64) so obtained is the model system for singular shocks, (3.2), identifying

$$v = \frac{1}{2}u^2 - \log \rho, \tag{7.81}$$

with symmetry groups shown in (7.9).

For all of these systems, the phase space coordinates obtained from (7.35) are

$$\begin{aligned} \alpha_1(w) &= v - \frac{1}{2}u^2 = -\log \rho \\ \alpha_2(w) &= -u. \end{aligned} \tag{7.82}$$

According to theorem 7.1, the Hugoniot locus for each of the systems (3.7), (3.2), and (7.75) should be translation invariant, using these coordinates. Indeed, denoting by [ ] the jump discontinuity of whatever quantity across an (ordinary) shock, for the system (3.7) the Hugoniot locus is described by

$$[\alpha_2]^2 = 4 \sinh^2([\alpha_1]/2). \tag{7.83}$$

For the system (3.2), the Hugoniot locus is compact, satisfying

$$[\alpha_2]^2 = [\alpha_1]^2 + \frac{1}{12}[\alpha_2]^4, \tag{7.84}$$

while for the system (7.75), the Hugoniot locus satisfies

$$[\alpha_2]^2 = 2[\alpha_1] \tanh\left([\alpha_1]/2\right). \tag{7.85}$$

CHAPTER 2

Delta-shocks

1. The structural conditions

Delta-shocks are typically identified with exceptional structural conditions in systems of conservation laws: real and equal characteristic speeds, a deficiency of corresponding eigenvectors, linearly degenerate characteristic families [B, ERS, S4, Z]. While the discussion and example of section 1 of the previous chapter shows that these conditions are not necessary, the structure of systems admitting delta-shock solutions is clearly related to the asymptotic behavior of the flux function $q(w)$ and the entropy density $U(w)$ as $|w| \to \infty$ within D.

While clarifying this, we obtain an unambiguous expression for the "fluid velocity" $u(w), w \in D$, permitting a more general definition of delta-shock solutions, avoiding an assumption such as $(3.18)_1$. (Here and throughout, subscripts on equation numbers will denote previous chapters.)

In view of the definition of a delta-shock given in section 4 of chapter 1, we identify delta-shocks with systems satisfying

$$|q(w)| + U(w) \leq c(1 + |w|), w \in D \tag{1.1}$$

for some constant c depending on the specific choice of D. For such systems, the possible values of singular mass in delta-shocks associated with single-valued delta-shock solutions may be restricted. In particular, denote by

$$\mathcal{M} = \{M \in \mathbb{R}^n \setminus \{0\} | M/\varepsilon \in D, \varepsilon > 0 \quad \text{sufficiently small}\} \tag{1.2}$$

and such that there exists a sequence of functions

$$\omega_M^\varepsilon : (-1, 1) \to \mathbb{R}_+, \varepsilon > 0, \tag{1.3}$$

with the following properties:

$$M\omega_M^\varepsilon(x) \in D, \quad x \in (-1, 1), \ \varepsilon > 0; \tag{1.4}$$

$$\int_{-1}^{1} \omega_M^\varepsilon(x) dx = 1; \tag{1.5}$$

$$\omega_M^\varepsilon \rightharpoonup \delta(x) \quad \text{as } \varepsilon \downarrow 0, \text{ weakly in the space of measures;} \tag{1.6}$$

$$\int_{-1}^{1} U(M\omega_M^\varepsilon(x)) dx \leq c, \text{ uniformly with respect to } \varepsilon, c \quad \text{depending on } M; \tag{1.7}$$

$$q(M\omega_M^\varepsilon) \rightharpoonup cM\delta(x) \text{ as } \varepsilon \downarrow 0, \quad \text{weakly in the space of} \tag{1.8}$$

measures, for some $c \in \mathbb{R}$ which may depend on M and on the specific sequence ω_M^ε.

The strong conditions on the elements of $\mathcal{M}$ are (1.4) and (1.8), required by theorem 3.3 of chapter 1. In particular $\mathcal{M}$ is isomorphic to D for systems such as (1.16), and (3.1), but not for the pair (4.28), for which $\mathcal{M} = \{(\rho, 0),\ \rho > 0\}$.

THEOREM 1.1. *Assume that the entropy density U is convex in D and satisfies*

$$U(w) \geq |w| \quad \text{for all } w \in D. \tag{1.9}$$

Then for all $M \in \mathcal{M}$

$$\begin{aligned} \bar{U}(M) &\stackrel{def}{=} \lim_{\varepsilon \downarrow 0} \varepsilon U(M/\varepsilon) \\ &\geq |M|. \end{aligned} \tag{1.10}$$

REMARKS. U does not have to be strictly convex.

The condition (1.9) is equivalent to $(3.10)_1$, as a linear term in w may be added to U as needed.

If U extends as convex to the convex hull of D, then $\bar{U}$ extends as convex and homogeneous of degree one to a convex cone $\bar{\mathcal{M}}$ containing $\mathcal{M}$. This will be assumed throughout.

The function $\bar{U}$ is the same as appearing in theorem 5.2 of chapter 1.

PROOF. From (1.5), (1.6) there exists $\delta_\varepsilon \downarrow 0$ as $\varepsilon \downarrow 0$ such that

$$\int_{-\delta_\varepsilon/2}^{\delta_\varepsilon/2} \omega_M^\varepsilon(x)dx \geq 1 - \delta_\varepsilon^2. \tag{1.11}$$

From (1.7), using the convexity of U,

$$\begin{aligned} \int_{-\delta_\varepsilon/2}^{\delta_\varepsilon/2} U(M\omega_M^\varepsilon(x))dx &\geq \int_{-\delta_e/2}^{\delta_\varepsilon/2} U(\frac{M}{\delta_\varepsilon} \int_{-\delta_\varepsilon/2}^{\delta_\varepsilon/2} \omega_M^\varepsilon(y)dy)dx \\ &= \delta_\varepsilon U(\frac{M}{\delta_e}) + O(\delta_\varepsilon) \end{aligned} \tag{1.12}$$

using (1.11). The left side of (1.12) is bounded uniformly with respect to ε from (1.7), so as $\varepsilon \downarrow 0$, a subsequence of $\{\delta_\varepsilon U(M/\delta_\varepsilon)\}$ converges. Indeed, from the assumed convexity of U, the entire sequence converges, so the limit in (1.10) is defined. Finally the lower bound (1.10) follows immediately from (1.9) and (1.12). □

The symmetry at infinity for "delta-shocks systems" depends on U, q satisfying slightly stronger assumptions than implied by (1.7), (1.8) respectively.

THEOREM 1.2. *Assume that U_w, q_w are sums of terms homogeneous of degree zero in w and terms vanishing in the limit $|w| \to \infty$. Then for all $M \in \mathcal{M}$*

$$\bar{F}(M) \stackrel{def}{=} \lim_{\varepsilon \downarrow 0} \varepsilon F(M/\varepsilon) \tag{1.13}$$

$$\bar{q}(M) \stackrel{def}{=} \lim_{\varepsilon \downarrow 0} \varepsilon q(M/\varepsilon) \tag{1.14}$$

exist, and at every interior point $M \in \mathcal{M}$

$$\bar{F}_M(M) = \bar{U}_M(M)\bar{q}_M(M). \tag{1.15}$$

Furthermore, there exists a smooth function $u : \mathcal{M} \to \mathbb{R}$ *such that for all* $M \in \mathcal{M}$

$$\bar{F}(M) = u(M)\bar{U}(M) \tag{1.16}$$

and

$$\bar{q}(M) = u(M)M. \tag{1.17}$$

Finally, $\bar{F}, \bar{q}$ *extend to* $\bar{\mathcal{M}}$ *as homogeneous of degree one and* u *as homogeneous of degree zero.*

REMARKS. In (1.15) subscripts M denote differentiation. In this case the constant c in (1.8) is independent of the sequence $\{w_M^\varepsilon\}$ and depends only on M. For a single-valued delta-shock solution, given a delta-shock of singular mass $M_i(t)$ with location $x_i(t)$, from $(3.57)_1$, (1.14) and (1.17), the speed of propagation is

$$x_{i,t}(t) = u(M_i(t)) \tag{1.18}$$

for almost all t, and from $(5.12)_1$, and (1.10), the singular entropy is

$$U_i(t) = \bar{U}(M_i(t)). \tag{1.19}$$

PROOF. The existence of $\bar{q}$ satisfying (1.14) follows from the assumption on q_w. Indeed, by assumption $\bar{U}_M(M), \bar{q}_M(M)$ are the terms homogeneous of degree zero in $U_w(M/\varepsilon), q_w(M/\varepsilon)$, for $\varepsilon > 0$ such that $M/\varepsilon \in D$, for M an interior point in $\mathcal{M}$. Then for M an interior point in $\mathcal{M}$ and $\varepsilon > 0$ such that $M/\varepsilon \in D$,

$$\begin{aligned} F_W(M/\varepsilon) &= U_w(M/\varepsilon)q_w(M/\varepsilon) \\ &= \bar{U}_M(M)\bar{q}_M(M) + o(1) \quad \text{as } |w| \to \infty \end{aligned} \tag{1.20}$$

so $\bar{F}$ determined from (1.13) satisfies (1.15).

Now for $M \in \mathcal{M}, u(M)$ is determined from (1.16), as $\bar{U}(M) > 0$ from (1.10); $\bar{q}, \bar{F}$ extend to $\bar{\mathcal{M}}$ as homogeneous of degree one and u as homogeneous of degree zero. Then from (1.8), it follows that $\bar{q}(M)$ is a scalar multiple of M for all $M \in \mathcal{M}$.

Differentiating (1.16) and using (1.15) we find

$$\bar{U}_M(M)\bar{q}_M(M) = \bar{U}(M)u_M(M) + u(M)\bar{U}_M(M). \tag{1.21}$$

Multiplying (1.21) from the right by M, using the homogeneity of $\bar{q}, u$, we obtain

$$\bar{U}_M(M)(\bar{q}(M) - u(M)M) = 0. \tag{1.22}$$

From the homogeneity of $\bar{U}$ and (1.10), we have $\bar{U}_M(M)M = \bar{U}(M) > 0$, so (1.22) suffices to prove (1.17). □

The familiar structural conditions for systems admitting delta-shock solutions result from special conditions on the lower order terms in U, q.

THEOREM 1.3. *Assume that for all* $w \in D$

$$U_{ww}(w) \geq 0 \tag{1.23}$$

and

$$\dim \ker U_{ww}(w) = 1. \tag{1.24}$$

Then the characteristic speeds $\lambda_i(w), i = 1, \ldots, n$ are real. At each $w \in D$, either there is a deficiency of corresponding eigenvectors, or else $\ker U_{ww}(w)$ is a right eigenvector of $q_w(w)$ and there exists a Friedrichs symmetrizer for the system at the point w.

PROOF. For $w \in D, \lambda_i(w)$ a characteristic speed and $r_i(w)$ the corresponding eigenvector

$$q_w(w)r_i(w) = \lambda_i(w)r_i(w) \tag{1.25}$$

we have [FL]

$$U_{ww}(w)q_w(w)r_i(w) = \lambda_i(w)U_{ww}(w)r_i(w) \tag{1.26}$$

with $U_{ww}(w)q_w$ symmetric. Thus for $r_i(w) \in \mathbb{C}^n \setminus \{0\}$,

$$r_i^\dagger(w)U_{ww}(w)q_w(w)r_i(w),\ r_i^\dagger(w)U_{ww}(w)r_i(w) \in \mathbb{R} \tag{1.27}$$

and if $\lambda_i(w) \notin \mathbb{R}$,

$$r_i^\dagger(w)U_{ww}(w)q_w(w)r_i(w) = \lambda_i(w)r_i^\dagger(w)U_{ww}(w)r_i(w) \tag{1.28}$$

can hold only with both sides equal to zero. But $\bar{\lambda}_i(w)$ is then also a (nonzero) characteristic speed with corresponding eigenvector $\bar{r}_i(w)$, so analogously with (1.28)

$$\bar{r}_i^\dagger(w)U_{ww}(w)q_w(w)\bar{r}_i(w) = \bar{r}_i^\dagger(w)U_{ww}(w)\bar{r}_i(w) = 0, \tag{1.29}$$

which is incompatible with (1.23) and (1.24).

Next suppose that at some point w there exist n linearly independent eigenvectors $r_1, \ldots, r_n$. Using (1.26), they are chosen, without loss of generality, so that

$$r_i^\dagger U_{ww} r_j = 0,\ i \notin j. \tag{1.30}$$

Using the symmetry of $U_{ww}q_w$, the transpose of (1.26) is

$$r_i^\dagger U_{ww} q_w = \lambda_i r_i^\dagger U_{ww}, \tag{1.31}$$

so for $U_{ww}r_i$ nonvanishing,

$$\ell_i = r_i^\dagger U_{ww},\ i = 1, \ldots, n \tag{1.32}$$

are left eigenvectors, just as when U is strictly convex.

Now from (1.24), the $\ell_i, i = 1, \ldots, n$ cannot be linearly independent; there exist $c_i \in \mathbb{R}$ not all zero such that

$$\sum_{i=1}^{n} c_i \ell_i = 0 \tag{1.33}$$

From (1.30) and (1.32)

$$\ell_i r_j = 0,\ i \neq j, \tag{1.34}$$

and since the r_j are linearly independent $\ell_i r_i = 0$ only if $\ell_i = 0$. Thus (1.33) is equivalent to

$$c_i \ell_i r_i = 0,\ i = 1, \ldots, n \tag{1.35}$$

which can hold only by a choice $c_1 = \ldots c_{n-1} = 0,\ c_n \neq 0, \ell_n = 0$ rearranging the eigenvectors if necessary. Thus

$$\ell_n^\dagger = U_{ww} r_n = 0 \tag{1.36}$$

and $\ker U_{ww}$ is one of the eigenvectors as claimed.

However, if there exists a full set of right eigenvectors, there also exists a full set of left eigenvectors, so there exists $\ell'_n \in \mathbb{R}^n/\{0\}$ satisfying

$$\ell'_n q_w = \lambda_n q_w, \ell'_n r_n = 1, \ell'_n r_i = 0, \ i = 1, \dots, n-1. \tag{1.37}$$

Then

$$\mathcal{F} = U_{ww} + \ell'^{\dagger}_n \ell'_n \tag{1.38}$$

is symmetric and positive definite, using (1.23), (1.24), (1.36), (1.37).

Finally

$$\begin{aligned} \mathcal{F} q_w &= U_{ww} q_w + \ell'^{\dagger}_n \ell'_n q_w \\ &= U_{ww} q_w + \lambda_n \ell'^{\dagger}_n \ell'_n \end{aligned} \tag{1.39}$$

is symmetric, using (1.37), so $\mathcal{F}$ is a Friedrichs symmetrizer as claimed. □

Examples of such systems satisfying (1.23) and (1.24) with complete sets of eigenvectors include the pairs (4.28), with U depending only on u. These systems generally do not admit strictly convex entropy densities.

THEOREM 1.4. *Assume $\mathcal{M}$ isomorphic to D, and q of the form*

$$q(w) = u(w)w, \ w \in D, \tag{1.40}$$

with u homogeneous of degree zero and $|u_w(w)| \neq 0$ for all $w \in D$. Then $u(w)$ is a characteristic speed of multiplicity n for all $w \in D$, to which there correspond $n-1$ linearly independent eigenvectors. The corresponding characteristic families are linearly degenerate.

PROOF. From (1.40)

$$\begin{aligned} q_w r &= ur + w(u_w r) \\ &= \lambda r \end{aligned} \tag{1.41}$$

has $n-1$ linearly independent solutions $u_w r = 0, \lambda = u$, so it suffices to prove that there are no other eigenvalues. Writing (1.41) as

$$(\lambda - u) r = w(u_w r) \tag{1.42}$$

for $\lambda \neq u, u_w r \neq 0$ and r is a scalar multiple of w. But $u_w w = 0$ by the homogeneity of u. □

The condition (1.40) is highly restrictive, but includes the case of zero-pressure gas dynamics $(3.1)_1$. More generally, theorem 1.2 is satisfied with q of the form

$$q(w) = u(w)w + b(w) \tag{1.43}$$

where

$$|b_w(w)| \to 0 \quad \text{as } |w| \to \infty. \tag{1.44}$$

A weaker condition than (1.40) leading to similar conclusions follows.

THEOREM 1.5. *Assume $\mathcal{M}$ isomorphic to D, that U, F are both homogeneous of degree one, and that q is of the form (1.43) with u obtained from (1.16) and b both homogeneous of degree zero. Assume in addition that for all $w \in D$,*

$$u_w(w) b_w(w) = 0. \tag{1.45}$$

Then for each $w \in D$, $u(w)$ is a characteristic speed of multiplicity at least two. There is a deficiency of eigenvectors corresponding to this characteristic speed, and all corresponding characteristic families are linearly degenerate.

PROOF. Theorem 1.2 holds with U and $\bar{U}, F$ and $\bar{F}$ coinciding, so u is homogeneous of degree zero and satisfies

$$F(w) = u(w)U(w) \tag{1.46}$$

throughout D. Differentiating (1.46) and using (1.43)

$$\begin{aligned} F_w &= Uu_w + uU_w \\ &= U_w q_w \\ &= U_w(u + wu_w + b_w) \\ &= uU_w + Uu_w + U_w b_w \end{aligned}$$

using the homogeneity of U in the last step. Using (1.46) we find

$$U_w b_w = 0 \tag{1.47}$$

and

$$U_w(q_w - u) = Uu_w. \tag{1.48}$$

From (1.45) and the homogeneity of u,

$$\begin{aligned} u_w q_w &= u_w(wu_w + u + b_w) \\ &= uu_w, \end{aligned} \tag{1.49}$$

so u_w is a left eigenvector of q_w corresponding to u.

From (1.48), since U satisfying (1.9) is not zero, U_w is a generalized left eigenvector corresponding to u, so u is of multiplicity at least two and q_w is not diagonizable. Thus the eigenvector deficiency.

A right eigenvector corresponding to u is w; the corresponding characteristic family is linearly degenerate, as $u_w w = 0$ by homogeneity of u.

Suppose there exists another linearly independent right eigenvector r, satisfying

$$q_w r = ur \tag{1.50}$$

and

$$u_w r \neq 0. \tag{1.51}$$

Without loss of generality we assume

$$U_w r = 0, \tag{1.52}$$

adding a multiple of w to r if necessary.

The proof now resembles the more general argument that multiple eigenvectors correspond to linearly degenerate characteristic families [D]. Differentiating (1.50) and multiplying by w

$$\begin{aligned} q_{ww} rw + q_w r_w w &= (u_w w)r + ur_w w \\ &= ur_w w, \end{aligned} \tag{1.53}$$

while differentiating $q_w w = uw$ and multiplying by r gives

$$q_{ww} wr + q_w r = (u_w r)w + ur, \tag{1.54}$$

so subtracting (1.58) from (1.53), we find

$$(q_w - u)(r_w w - r) = (u_w r)w. \tag{1.55}$$

Consider the restriction of q_w to the space spanned by $\{w, r_w w - r, r\}$. With respect to this basis, q_w assumes the form

$$\begin{pmatrix} u & u_w r & 0 \\ 0 & u & 0 \\ 0 & 0 & u \end{pmatrix}$$

using (1.50) and (1.55).

Next consider the representations of U_w and u_w with respect to this basis. Determined from (1.48), (1.49) and (1.52) with q_w as shown above we identify unambiguously

$$U_w = (U \quad 0 \quad 0) \tag{1.56}$$

$$u_w = (0 \quad U/u_w r \quad 0). \tag{1.57}$$

But then from (1.57) we have $u_w r = 0$, contradicting (1.51) and thus proving the claim of linearly degenerate corresponding characteristic families. □

2. Definition of solutions

In view of the results of the preceding section in general and the conclusion of theorem 1.2 in particular, we consider here systems of the form

$$w_t + (u(w)w + b(w))_x = 0, \; b(w) = b_0(w) + b_1(w), \tag{2.1}$$

with u, b_0 smooth and homogeneous of degree zero with respect to $w \in D$, and b_1 smooth and satisfying

$$|b_1(w)| \le o(1), \; |b_{1,w}(w)| \le o(1/|w|) \quad \text{as } |w| \to \infty. \tag{2.2}$$

The explicit appearance of the "fluid velocity" $u(w)$ in (2.1) allows a broader and simpler definition of a distribution solution than that given in section 4 of the preceding chapter.

DEFINITION. **A distribution solution of a system (2.1) is a pair** $(X, \hat{w}), X \in W^{1,\infty} : \mathbb{R} \times \mathbb{R}_+ \to \mathbb{R}, \; \hat{w}(\cdot, t) \in BV : \mathbb{R} \to D, \; t > 0$ **satisfying**

$$X_y(y,t) \ge 0, \; X(\pm\infty, t) = \pm\infty, \tag{2.3}$$

$$X_t(y,t) = u(\hat{w}(y,t)), \; \textbf{a.e. in } \mathbb{R} \times \mathbb{R}_+ \tag{2.4}$$

and

$$\int_{\mathbb{R}_+}\int_{\mathbb{R}} [\hat{w}(y,t) \cdot \theta(X(y,t),t)_t + (b_0(\hat{w}(y,t)) + b_1(\hat{w}(y,t)/X_y(y,t))) \cdot \theta(X(y,t),t)_y] dydt = 0 \tag{2.5}$$

for all $\theta \in C_0 : \mathbb{R} \times \mathbb{R}_+ \to \mathbb{R}^n$, **where** $b_1(\hat{w}/X_y) \overset{\text{def}}{=} 0$ **where** $X_y = 0$.

Given such $(X, \hat{w})$, we recover the corresponding measure $w(\cdot, t)$ on $\mathbb{R}$, pointwise in t, from

$$\int_{\mathbb{R}} w(x,t) \cdot \phi(x) dx \overset{\text{def}}{=} \int_{\mathbb{R}} \hat{w}(y,t)\phi(X(y,t))dy \tag{2.6}$$

for any $\phi \in C_0 : \mathbb{R} \to \mathbb{R}^n$.

THEOREM 2.1. *Given such* $(X, \hat{w})$, *there exists a sequence* $w^\varepsilon : \mathbb{R} \times \mathbb{R}_+ \to D, \varepsilon > 0$, *such that*

$$w^\varepsilon(\cdot, t) \in L^1_{loc} : \mathbb{R} \to D \quad \text{uniformly with respect to } \varepsilon, \tag{2.7}$$

pointwise in t;

$$w^\varepsilon(\cdot, t) \rightharpoonup w(\cdot, t) \quad \text{as } \varepsilon \downarrow 0, \quad \text{weakly in the space} \tag{2.8}$$

of measures on $\mathbb{R}$, *pointwise in* t;

$$w^\varepsilon_t + (u(w^\varepsilon)w^\varepsilon + b(w^\varepsilon))_x \rightharpoonup 0 \quad \text{as } \varepsilon \downarrow 0, \tag{2.9}$$

weakly in the sense of distributions.

PROOF. Determine $X^\varepsilon, w^\varepsilon$ from

$$X^\varepsilon(y,t) = X(y,t) + \varepsilon y, \tag{2.10}$$

$$\begin{aligned} w^\varepsilon(X^\varepsilon(y,t),t) &= \hat{w}(y,t)/X^\varepsilon_y(y,t) \\ &= \hat{w}(y,t)/(\varepsilon + X_y(y,t)), \ y \in \mathbb{R}, \ t \geq 0. \end{aligned} \tag{2.11}$$

Then for $\phi \in C_0 : \mathbb{R} \to \mathbb{R}^n$, using (2.1), then (2.10)

$$\begin{aligned} \int_{\mathbb{R}} w^\varepsilon(x,t) \cdot \phi(x) dx &= \int_{\mathbb{R}} w^\varepsilon(X^\varepsilon(\cdot,t),t) \cdot \phi(X^\varepsilon(\cdot,t)) dX^\varepsilon(\cdot,t) \\ &= \int_{\mathbb{R}} \hat{w}(y,t) \cdot \phi(X^\varepsilon(y,t)) dy \\ &= \int_{\mathbb{R}} \hat{w}(y,t) \cdot \phi(X(y,t) + \varepsilon y) dy \\ &= \int_{\mathbb{R}} w(x) \cdot \phi(x) dx + \int_{\mathbb{R}} \hat{w}(y,t) \cdot (\phi(X(y,t) + \varepsilon y) \\ &\quad - \phi(X(y,t))) dy \end{aligned} \tag{2.12}$$

recalling (2.6) in the last step. Now (2.7) and (2.8) follow from (2.12).

For $\theta \in C^1_0 : \mathbb{R} \times \mathbb{R}_+ \to \mathbb{R}^n$

$$\begin{aligned} &\iint [w^\varepsilon(x,t) \cdot \theta_t(x,t) + (u(w^\varepsilon(x,t))w^\varepsilon(x,t) + b(w^\varepsilon(x,t))) \cdot \theta_x(x,t)] dx dt \\ &= \iint [\hat{w}(y,t) \cdot \theta_t(X^\varepsilon(y,t),t) + (u(w^\varepsilon(y,t))w^\varepsilon(y,t) + b(w^\varepsilon(y,t))) \\ &\qquad \cdot \theta(X^\varepsilon(y,t),t)_y] dy dt, \end{aligned}$$

(abbreviating $w^\varepsilon(X^\varepsilon(y,t),t) = w^\varepsilon(y,t)$ here and below),

$$\begin{aligned} = \iint [\hat{w}(y,t) \cdot \theta(X^\varepsilon(y,t),t)_t + ((u(w^\varepsilon(y,t)) - X^\varepsilon_t(y,t))w^\varepsilon(y,t) \\ + b(w^\varepsilon(y,t))) \cdot \theta(X^\varepsilon(y,t),t)_y] dy dt \end{aligned} \tag{2.13}$$

as in the first step in obtaining (2.12), with ϕ replaced by θ_t. We can eliminate one term in (2.13) using (2.10), then (2.4), then the homogeneity of u and (2.11), to obtain

$$
\begin{aligned}
X_t^\varepsilon(y,t) &= X_t(y,t)\\
&= u(\hat{w}(y,t))\\
&= u(w^\varepsilon(y,t)).
\end{aligned}
\tag{2.14}
$$

Now using (2.14), (2.13) becomes

$$
\begin{aligned}
&\iint [\hat{w}(y,t)\cdot(\theta(X^\varepsilon(y,t),t)) - \theta(X(y,t),t))_t\\
&\quad + b(w^\varepsilon(y,t))\cdot\theta(X^\varepsilon(y,t),t)_y - (b_0(\hat{w}(y,t)) + b_1(\hat{w}(y,t)/X_y(y,t)))\\
&\quad\cdot\theta(X(y,t),t)_y]dydt\\
&= \iint \hat{w}(y,t)\cdot(\theta_t(X^\varepsilon(y,t),t) - \theta_t(X(y,t),t))dydt\\
&\quad + \iint \hat{w}(y,t)\cdot(\theta_x(X^\varepsilon(y,t),t) - \theta_x(X(y,t),t))X_t(y,t)dydt\\
&\quad + \iint b_0(\hat{w}(y,t))\cdot(\theta(X^\varepsilon(y,t),t) - \theta(X(y,t),t))_y dydt\\
&\quad + \iint (b_1(\hat{w}(y,t)/(X_y(y,t)+\varepsilon)\cdot\theta(X^\varepsilon(y,t),t)_y\\
&\qquad - b_1(\hat{w}(y,t)/X_y(y,t)\cdot\theta(X(y,t),t)_y)dydt
\end{aligned}
\tag{2.15}
$$

using the homogeneity of b_0 and (2.11) in the third term. Using the regularity of θ and the assumed conditions (2.2) on b_1, all of the right-hand terms in (2.15) vanish in the limit $\varepsilon \downarrow 0$. □

COROLLARY. *Assume $(X, \hat{w})$ is a distribution solution and that $X_y(y,t) > 0$; then w obtained from*

$$
w(X(y,t),t) = \hat{w}(y,t)/X_y(y,t)
\tag{2.16}
$$

is a weak solution of (2.1).

PROOF. In this case one may pass to the limit $\varepsilon = 0$ in (2.10), (2.11). □

LEMMA 2.2. *Assume D a cone in $\mathbb{R}^n$ and a system of the form (2.1) equipped with an entropy density/ flux $U(w), F(w)$, such that $U_w(w)$ is a term homogeneous of degree zero in w plus a term vanishing in the limit $|w| \to \infty$. Then D is isomorphic to $\mathcal{M}$, the assumptions of theorem 1.2 hold, and $\bar{F}(w) = u(w)\bar{U}(w)$.*

PROOF. Given $q(w) = u(w)w + b(w)$ and U as assumed, (1.1) holds, and for any $M \in D, \omega_M^\varepsilon$ satisfying (1.3, 1.5, 1.6), it follows that (1.4) and (1.7) hold. With $q(w)$ of this form (1.8) holds with the constant c equal to $u(M)$. Thus $\mathcal{M}$ is isomorphic to D.

Furthermore

$$
q_w(w) = u(w)I_{n\times n} + wu_w(w) + b_w(w)
\tag{2.17}
$$

the first two right-hand terms homogeneous of degree zero and the last term vanishing in the limit as $|w| \to \infty$. So theorem 1.2 applies with u in (2.1) the same function as appearing in (1.17). □

THEOREM 2.3. *For a system of the form (2.1) also satisfying $(3.18)_1$, the definitions of distribution solutions given here and in section 4, chapter 1, are equivalent.*

PROOF. Using $(4.6)_1$ we identify

$$X_y(y,t) = v(y,t) = 1/a \cdot w(X(y,t),t), \tag{2.18}$$

so $(4.7)_1$ and (2.16) show that the values of

$$w(X(y,t),t) = \hat{w}(y,t)/v(y,t) \tag{2.19}$$

coincide.

For such a system

$$\Sigma = \{(\frac{1}{a \cdot w}, \frac{w}{a \cdot w}, \frac{u(w)w + b(w)}{a \cdot w}), w \in D\} \tag{2.20}$$

and

$$\Sigma_0 = \{(0, \frac{w}{a \cdot w}, \frac{u(w)w}{a \cdot w}), \quad w \in D\}. \tag{2.21}$$

The left side of $(4.9)_1$ is now, using (2.18), (2.19), (2.20), (2.21), (2.16), (2.4), and noting that $a \cdot \hat{w}(y,t) = 1$,

$$\begin{aligned}
&\iint \hat{w}(y,t) \cdot (\theta(X(y,t),t)_t - \theta_x(X(y,t),t)X_t(y,t))dydt \\
&+ \iint_{X_y>0} [u(\hat{w}(y,t))\hat{w}(y,t) + v(y,t)(b_0(\hat{w}(y,t)) + b_1(\hat{w}(y,t)/v(y,t)))] \\
&\qquad \cdot \theta_x(X(y,t),t)dydt \\
&+ \iint_{X_y=0} u(\hat{w}(y,t))\hat{w}(y,t) \cdot \theta_x(X(y,t),t)dydt \\
&= \iint \hat{w}(y,t) \cdot \theta(X(y,t),t)_t dydt \\
&+ \iint_{X_y>0} (b_0(\hat{w}(y,t)) + b_1(\hat{w}(y,t)/X_y(y,t)) \cdot \theta(X(y,t),t)_y dydt
\end{aligned} \tag{2.22}$$

which coincides with the left side of (2.5). ☐

THEOREM 2.4. *Assume a distribution solution* $(X, \hat{w})$ *with*

$$\{X_y = 0\} = \bigcup_i \{(y,t), y_{i-}(t) < y < y_{i+}(t), \ t \in I_i\} \tag{2.23}$$

possibly excepting a set of measure zero in $\mathbb{R} \times \mathbb{R}_+$. *Then* $(X, \hat{w})$ *determines a delta-shock solution as defined in section 4, chapter 1, with*

$$\tilde{w}(X(y,t),t) = \hat{w}(y,t)/X_y(y,t), \quad X_y(y,t) > 0 \tag{2.24}$$

$$M_i(t) = \int_{y_{i-}(t)}^{y_{i+}(t)} \hat{w}(y,t)dy \tag{2.25}$$

$$x_{i,t}(t) = u(\hat{w}(y,t)), \ y_{i-}(t) < y < y_{i+}(t). \tag{2.26}$$

PROOF. The identification of the singular mass $M_i(t)$ (2.25) follows from (2.6). From (2.4), $u(\hat{w}(y,t))$ must be independent of y in the intervals $(y_{i-}(t), y_{i+}(t))$, so (2.26) follows from $(3.57)_1$ using $q = uw + b$. Now $(3.20)_1$ holds with $\tilde{w}$ obtained from (2.24). Then $q(w)$ of the form $(3.57)_1$ follows by application of theorem 2.3. ☐

COROLLARY. *Assume in addition that $\hat{w}(y,t)$, is independent of y in each interval $(y_{i-}(t), y_{i+}(t))$. Then the delta-shock solution is single-valued, with singular entropy $U_i(t)$ as given in (1.19).*

PROOF. This follows easily by application of theorem 2.3. □

3. Delta-shocks and Lagrangian coordinates

Juxtaposing V with X_y and W with $\hat{w}$, distribution solutions $(X, \hat{w})$ of (2.1) resemble ordinary weak solutions (V, W) of the system

$$V_t - u(W)_y = 0, \tag{3.1}$$

$$W_t + (b_0(W) + b_1(W/V))_y = 0. \tag{3.2}$$

There are two important differences. Negative values of V are permitted in (3.1), (3.2), requiring a suitable extension of b_1. As against that, there are fewer eligible test functions for (2.5) than for (3.2), as $\theta(X(\cdot,t))$ is independent of y within intervals in which $X_y(\cdot,t)$ vanishes.

An alternative expression of the connection between (2.1) and (3.1), (3.2) is the observation that (3.1), (3.2) is just the Lagrangian form of an extended, uncoupled form of (3.1),

$$\rho_t + (\rho u)_x = 0, \tag{3.3}$$

$$w_t + (wu + b_0(w) + b_1(w))_x = 0, \tag{3.4}$$

identifying

$$V = 1/\rho, \ W = w/\rho \tag{3.5}$$

and using u, b_0 homogeneous of degree zero in w. The uncoupled form of (3.3), (3.4) (w satisfying (3.4) does not depend on ρ) is curiously masked in (3.1), (3.2).

A number of structural conditions for the system (3.1), (3.2) are thus inherited from those for (3.4) (or (2.1)).

THEOREM 3.1. *Assume that the system (3.4) is equipped with an entropy density/flux U, F, with U satisfying (1.23), (1.24). Then for $V > 0$, the characteristic speeds $\lambda_i^*(V,W), i = 1,\ldots,n+1$, are real and given by*

$$\lambda_i^*(V,W) = (\lambda_i(W/V) - u(W))/V, \ \ i = 1,\ldots,n; \lambda_{n+1}^* = 0, \tag{3.6}$$

where $\lambda_1(w),\ldots,\lambda_n(w)$ are the characteristic speeds for the system (2.1).

Furthermore, the system (3.1), (3.2) admits an entropy density / flux

$$U^*(V,W) = VU(W/V) \tag{3.7}$$

$$F^*(V,W) = F(W/V) - u(W)U(W/V) \tag{3.8}$$

with U^* convex in V, W.

PROOF. Using (3.5), the results (3.6), for $i = 1,\ldots,n$, (3.7), (3.8) are just the usual results of transformation to a Lagrangian space coordinate. An elementary calculation shows that affixing (3.3) to (3.4) simply adds a characteristic speed $\lambda_{n+1} = u$.

The (not strict) convexity of U^* follows easily from (1.23) and (3.7). □

Even with b_1 satisfying (2.2) the extension of (3.2) to nonpositive V is problematical. The following illustrates a possible such extension.

THEOREM 3.2. *Assume that U_w is homogeneous of degree zero plus a term vanishing in the limit $|w| \to \infty$, that*

$$U(W/V) - U_w(W/V) \cdot W/V \to 0 \tag{3.9}$$

and that

$$-b_{1,w}(W/V) \cdot W/V^2 \to b^* \tag{3.10}$$

as $V \downarrow 0$ uniformly for W in compact sets of D, for some $b^* \in \mathbb{R}^n$ independent of W. Then the extension

$$b_1(V,W) = b^* V \tag{3.11}$$

for $V \le 0$ extends the system (3.1), (3.2) to $V \le 0$ such that the characteristic speeds are continuous and satisfy

$$\lambda_i^*(V,W) = \lim_{V' \downarrow 0} \lambda_i^*(V',W), \quad V \le 0 \tag{3.12}$$

and the entropy density / flux is C^1 and satisfies

$$U^*(V,W) = \bar{U}(W), \tag{3.13}$$

$$F^*(V,W) = F_0(W) + \bar{U}_W(W) b^* V, \;\; V \le 0 \tag{3.14}$$

with

$$F_{0,W}(W) = \bar{U}_W(W) b_{0,W}(W). \tag{3.15}$$

PROOF. From (1.9) and (3.7), as $V \downarrow 0$

$$U^*(V,W) \to \bar{U}(W) \tag{3.16}$$

and by our assumption on U_W, differentiating (3.7) with respect to W

$$U_W^*(V,W) \to \bar{U}_W(W). \tag{3.17}$$

Differentiating (3.7) with respect to V, using (3.9),

$$U_V^*(V,W) \to 0. \tag{3.18}$$

Denote by

$$q^*(V,W) = \begin{pmatrix} -u(W) \\ b_0(W) + b_1(W/V) \end{pmatrix} \tag{3.19}$$

the flux vector associated with (3.1), (3.2). Then using (2.1) and (3.10)

$$\begin{aligned} \frac{\partial q^*(V,W)}{\partial(V,W)} &= \begin{pmatrix} 0 & -u_W(W) \\ \frac{-b_{1,w}W}{V^2} & b_{0,W}(W) + \frac{b_{1w}(W,V)}{V} \end{pmatrix} \\ &\to \begin{pmatrix} 0 & -u_W(W) \\ b^* & b_{0,W}(W) \end{pmatrix} \end{aligned} \tag{3.20}$$

as $V \downarrow 0$. The extension (3.11) of b_1 implies the extension of $\partial q^*(V,W)/\partial(V,W)$ to negative V as the limit shown in (3.20), independent of W. The characteristic speeds $\lambda_i^*(V,W)$, $i = 1, \dots, n+1$ are just the eigenvalues of $\partial q^*(V,W)/\partial(V,W)$ so they are continuous and become independent of V for V negative as claimed.

For $V > 0$, the entropy density / flux satisfies

$$(F_V^* \; F_W^*) = (U_V^* \; U_W^*) \frac{\partial q^*(V,w)}{\partial(V,W)}. \tag{3.21}$$

Taking the limit as $V \downarrow 0$, using (3.17), (3.18) we obtain

$$F^*_V(V,W) \to \bar{U}_W(W)b^* \tag{3.22}$$

and

$$F^*_W(U,W) \to \bar{U}_W(W)b_{0,W}(W). \tag{3.23}$$

Thus extending U^* as in (3.13) and F^* as in (3.14), (3.15), both U^*, F^* are C^1 functions of V, W and satisfy (3.21) for $V \le 0$ as needed. □

Given a weak solution of (3.1), (3.2), a simple procedure allows the recovery of a distribution solution, indeed a single-valued delta-shock solution, of (2.1), at least for some finite time after singularities form. We assume that V is initially positive, and remains positive for $|y|$ sufficiently large, but becomes negative on a set of finite measure after finite time. This results in the formation of singularities in the corresponding distribution solution of (2.1).

Specifically, at each t where $V(\cdot,t)$ attains negative values on a set of finite measure we determine points $y_{i\pm}(t)$, $i = 1, \cdots$, satisfying

$$y_{i-}(t) < y_{i+}(t) < y_{(i+1)-}(t) < \dots \tag{3.24}$$

and such that

$$\{y|V(y,t) < 0\} \subset \mathcal{S}(t) \stackrel{\text{def}}{=} \bigcup_{i:t\in I_i} \{(y_{i-}(t), y_{i+}(t))\}. \tag{3.25}$$

In addition we require

$$V(y_{i\pm}(t),t) \ge 0, \tag{3.26}$$

and

$$\int\limits_{y_{i-}(t)}^{y_{i+}(t)} V(y,t)dy = 0, \; i = 1, \dots. \tag{3.27}$$

Then for $y \notin \mathcal{S}(t)$

$$X_y(y,t) = V(y,t) \tag{3.28}$$

and

$$\hat{w}(y,t) = W(y,t). \tag{3.29}$$

Within each interval $y_{i-}(t) < y < y_{i+}(t)$

$$X(y,t) = X(y_{i\pm}(t) \pm 0, t) \stackrel{\text{def}}{=} x_i(t), \tag{3.30}$$

and

$$\hat{w}(y,t) = M_i(t)/(y_{i+}(t) - y_{i-}(t)), \tag{3.31}$$

where

$$M_i(t) \stackrel{\text{def}}{=} \int\limits_{y_{i-}(t)}^{y_{i+}(t)} W(y,t)dy. \tag{3.32}$$

The continuity of $X(\cdot,t)$ with respect to y follows from (3.27). Finally, the position functions $x_i(t)$ are required to satisfy

$$x_{i,t}(t) = u(M_i(t)) \tag{3.33}$$

for almost all $t \in I_i$.This is the problematical condition, as it is not clear a priori that the $y_{i\pm}(t)$ can be continued in time such that (3.33) is satisfied.

THEOREM 3.3. *$(X, \hat{w})$ determined from (3.24)-(3.33) determine a distribution solution of (2.1).*

PROOF. Clearly $X_y \geq 0$; from (3.1) and (3.28) it follows that (2.4) is satisfied for almost all $(y,t), y \notin \mathcal{S}(t)$. Within $\mathcal{S}(t)$, (2.4) follows from (3.31) and (3.33).

Finally (2.5) follows immediately from (3.2), (3.29) and (3.31), as $X_y(\cdot,t)$ and $\theta(X(\cdot,t),t)_y$ vanish within $\mathcal{S}(t)$, so the values of $b(\hat{w})$ in these intervals are immaterial. □

Given (V,W) the $(X,\hat{w})$ so determined is in general not unique; this will be discussed in the next section.

The problematical step in the determination of $(X,\hat{w})$ is the determination of the $y_{i\pm}(t)$ such that (3.30) and (3.26) hold.

From (3.28), (3.1), for almost all (y,t), $y \notin \mathcal{S}(t)$, we have

$$X_t(y,t) = u(W(y,t)). \tag{3.34}$$

Differentiating (3.30) with respect to t and using (3.28), (3.34) and (3.33) we obtain ordinary differential equations for $y_{i\pm}$,

$$V(y_{i\pm}(t) \pm 0, t)y_{i\pm,t} + u(W(y_{i\pm}(t) \pm 0, t)) = u(M_i(t)). \tag{3.35}$$

As it suffices that $y_{i\pm}$ satisfy (3.30), a weak solution of (3.35) is sufficient, but may not exist in a neighborhood where $V(y_{i\pm}(t) \pm 0, t) = 0$.

Indeed this occurs at a point y_1, t_1 where a singularity forms spontaneously. Assuming V, W smooth in a neighborhood of such a point, in the simplest, nonetheless generic case $V(\cdot,t_1)$ will have a strict local minimum at $y = y_1$, and $V(y_1,\cdot)$ will be decreasing at a finite rate at $t = t_1$, so we assume for simplicity

$$V(y_1,t_1) = 0 \tag{3.36}$$

$$V_y(y_1,t_1) = 0 \tag{3.37}$$

$$V_{yy}(y_1,t_1) > 0 \tag{3.38}$$

$$\begin{aligned} V_t(y_1,b_1) &= u(W(\cdot,t_1)_y \Big|_{y=y_1} \\ &< 0 \end{aligned} \tag{3.39}$$

using (3.1) at the point $y = y_1$, $t = t_1$ in (3.39).

Anticipating a weak solution of (3.35) in a neighborhood of $t = t_1$, we rewrite (3.35) in the form

$$\begin{aligned} &V(y_{i\pm}(t) \pm 0, t)\frac{d}{dt}\Big(\frac{(y_{i\pm}(t) - y_1)^2}{2}\Big) \\ &= -(y_{i\pm}(t) - y_1)(u(W(y_{i\pm}(t) \pm 0, t)) - u(M_i(t))). \end{aligned} \tag{3.40}$$

We make a Taylor expansion of (3.40), using (3.32) and the homogeneity of degree zero of u,

$$u(M_i(t)) = u\Big(\frac{1}{y_{i+}(t) - y_{i-}(t)} \int_{y_i-(t)}^{y_{i+}(t)} W(y,t)dy\Big)$$

$$= u(W(y_1,t)) + O(|y_{i+}(t) + y_{i-}(t) - 2y_1| + (y_{i+}(t) - y_{i-}(t))^2). \tag{3.41}$$

Using (3.36), (3.37), (3.38),

$$V(y_{i\pm}(t) \pm 0, t) = \frac{1}{2}(y_{i\pm}(t) - y_1)^2 V_{yy}(y_1, t_1) + O(y_{i\pm}(t) - y_1)^3 \tag{3.42}$$

and using (3.39) and (3.41)

$$u(W(y_{i\pm}(t) \pm 0, t)) - u(M_i(t)) = (y_{i\pm}(t) - y_1)V_t(y_1, t_1) + O(y_{i+}(t) - y_{i-}(t))^2 + O(|y_{i+}(t) + y_{i-}(t) - 2y_1|). \tag{3.43}$$

Inserting (3.42) and (3.43) into (3.41), we obtain

$$V_{yy}(y_1, t_1)\frac{d}{dt}(y_{i\pm}(t) - y_1)^2 = -\,4V_t(y_1, t_1) + O(y_{i+}(t) - y_{i-}(t)) + O(\frac{y_{i+}(t) + y_{i-}(t) - 2y_1}{y_{i+}(t) - y_{i-}(t)}), \tag{3.44}$$

$$y_{i\pm}(t) - y_1 = \pm\, 2(-V_t(y_1, b_1)/V_{yy}(y_1, t_1))^{1/2}(t - t_1)^{1/2} + O(t - t_1) \tag{3.45}$$

and we have obtained

THEOREM 3.4. *Given a weak solution (V, W) of (3.1), (3.2), assume that there are only a finite number of points (y_0, t_0) where a discontinuity may form spontaneously, i.e., where $V(\cdot, t)$ is nonnegative in a neighborhood of y_0 for $t \leq t_0 (t_0 - t$ small), but for any $t > t_0, V(\cdot, t)$ achieves negative values in any neighborhood of y_0. Let t_1 be the smallest value of t_0 for which such points exist, and assume that (3.37), (3.38), (3.39) hold at all points y_1 where (3.36) holds.*

Then for some finite time after t_1, there exists a single-valued delta-shock solution $(X, \hat{w})$ of (2.1).

The systems of conservation laws specifically identified with delta-shocks are distinguished by several features. Classical weak solutions doe not exist for the system (2.1) for initial value problems of interest, whereas the corresponding system (3.1), (3.2) is often, if not always, easy to solve. This (obviously imprecise) notion will be illustrated by several examples in subsequent sections.

In addition, such systems have specific structural features permitting the continuation of the constructed distribution solution indefinitely in time, sometimes uniquely. This question is addressed next.

4. Continuation of delta-shock solutions

The central question to which this chapter is directed is of course the existence and uniqueness of delta-shock solutions for systems of the form (2.1). Given a weak solution of the system (3.1), (3.2), the discussion of the preceding section reduces the question of existence of a delta-shock solution to that of determining the $y_{i\pm}(t), t \in I_i$, such that (3.26), (3.27), (3.30), and (3.33) hold. The main results of this chapter are two theorems giving sufficient conditions that this can be done indefinitely in time.

In both cases significant additional assumptions on the system (2.1) are required, but in both cases very simple systems satisfying these assumptions exist.

The question of uniqueness is more complicated. First, it is not clear that all delta-shock solutions of (2.1) are related to weak solutions of (3.1), (3.2). Second, the obtained solutions of (3.1), (3.2) presumably depend on how the function $b(W/V)$ is extended to V negative. Third, ordinary weak solutions of (3.1), (3.2) are in general not unique. Finally, the construction of a delta-shock solution from a given weak solution of (3.1), (3.2) is in general not unique. We shall present some partial results in this direction below. In a few cases, such as "zero-pressure gas dynamics" $(3.1)_1$, the uniqueness question can be completely resolved [B, CLZ, ERS, LW, LZ, TZ, TZZ].

Given a weak solution (V, W) of (3.1), (3.2), there exists a function

$$\underline{X} \in W^{1,\infty}(\mathbb{R} \times \mathbb{R}_+ \to \mathbb{R}) \tag{4.1}$$

satisfying

$$\underline{X}_y(y,t) = V(y,t) \tag{4.2}$$

$$\underline{X}_t(y,t) = u(W(y,t)) \tag{4.3}$$

$$\underline{X}(\pm\infty, t) = \pm\infty \tag{4.4}$$

for almost all $(y,t) \in \mathbb{R} \times \mathbb{R}_+$. The Lagrange function X identified with a distribution solution of (2.1) satisfies

$$X(y,t) = \underline{X}(y,t), \quad y \notin \mathcal{S}(t) \tag{4.5}$$

$$X(y,t) = \underline{X}(y_{i\pm}(t), t), \quad y \in [y_{i-}(t), y_{i+}(t)], \ t \in I_i \tag{4.6}$$

using (3.27) and (3.29), with $\mathcal{S}(t)$ given in (3.25). Thus the problem of constructing a delta-shock solution of (2.1) from a weak solution of (3.11), (3.2) is precisely that of finding the $y_{i\pm}(t)$, $t \in I_i$, such that (3.25), (3.26) hold, and

$$\underline{X}(y_{i\pm}(t), t) = x_i(t), \ t \in I_i \tag{4.7}$$

with the shock position $x_i(t)$ determined from (3.33).

For $(y_{i\pm}(t), t)$ such that

$$V(y_{i\pm}(t), t) > 0, \tag{4.8}$$

the $y_{i\pm}$ continue via the differential equation (3.35).

The conditions (3.36-3.39) suffice that the $y_{i\pm}(t)$ exist in a neighborhood of a point where delta-shocks from spontaneously.

The $y_{i\pm}$ do not continue at a point $(\hat{x}, \hat{t})$ where two or more delta-shocks collide. However, if

$$V(y_{j\pm}(\hat{t}) \pm 0, \ \hat{t} + 0) > 0 \tag{4.9}$$

where $x_j(t)$ denotes the position of the delta-shock resulting from the collision, (satisfying $x_j(\hat{t}+0) = \hat{x}$) then (3.35) holds in the sense of forward time derivatives, $y_{j\pm,t}(\hat{t}+0)$ exist and the delta-shock solution can be continued.

Thus the danger to continuation of a delta-shock solution is that at some t_0 the function $\underline{X}(\cdot, t_0)$ attains a local extremum at one of the points $y_{i\pm}(t_0)$, $t_0 \in I_i$. However, as the regions in which $\underline{X}(\cdot, t_0 - \varepsilon)$, $\varepsilon > 0$, is nonincreasing with respect to y are already within the set $\mathcal{S}(t_0 - \varepsilon)$, the attaining of a local maximum of $\underline{X}(\cdot, t_0)$ at $y_{i+}(t_0)$ or a local minimum at $y_{i-}(t_0)$ necessarily implies a collision of delta shocks and can be dismissed. Thus the dangerous condition is that $\underline{X}(\cdot, t_0)$ attains a local minimum (respectively, a local maximum) at $y_{i+}(t_0)$ (respectively, $y_{i-}(t_0)$).

THEOREM 4.1. *Assume that the function $u(w)$ is such that*

$$\int_{y_{i-}(t)}^{y_{i+}(t)} u(w(y,t))dy = (y_{i+}(t) - y_{i-}(t))u(M_i(t)) \tag{4.10}$$

with $M_i(t)$ as in (3.32). Then given an ordinary weak solution (V, W) of (3.1), (3.2), a delta-shock solution of (2.1) can be continued indefinitely in time.

REMARKS. Uniqueness is not true in this generality, even for the given (V, W).

No special assumptions are needed about the spontaneous formation of delta-shocks.

The relation (4.10) would follow if u were linear in W, but this is incompatible with u homogeneous of degree zero. Nevertheless, (4.1) is true for Euler systems, where $u(W) = W_2/W_1$, $w_1 > 0$ and the first component of b vanishes identically. Then it is no loss of generality to choose $\rho(\cdot, 0) = w_1(\cdot, 0)$ in (3.3), (3.4), so that $\rho(x,t) = w_1(x,t)$, so $W_1(y,t) = w_1(x,t)/\rho(x,t) = 1$. With W_1 so restricted, $u(W) = W_2$ and (4.10) follows.

PROOF. Given (4.10) the constructed Lagrange function X will satisfy

$$\int_{y_{i-}(t)}^{y_{i+}(t)} (\underline{X}(y,t) - X(y,t))dy = 0. \tag{4.11}$$

By definition, (4.11) will hold at points where delta-shocks form spontaneously. Using (4.6), (4.3), (3.33)

$$\begin{aligned} \frac{d}{dt} \int_{y_{i-}(t)}^{y_{i+}(t)} (\underline{X}(y,t) - X(y,t))dy \\ &= \int_{y_{i-}(t)}^{y_{i+}(t)} (u(W(y,t)) - u(M_i(t)))dt \\ &= 0. \end{aligned}$$

Finally, if two w more delta-shocks collide, each satisfying (4.11), then (4.11) will be satisfied initially by the resulting delta-shock. Thus (4.11) holds for all time.

We observe that a partial integration of (4.11), using (4.6), gives

$$\int_{y_{i-}(t)}^{y_{i+}(t)} yV(y,t)dy = 0. \tag{4.12}$$

Thus when (4.10) holds, the $y_{\pm i}(t)$ can be determined from (3.27) and (4.12), given $V(\cdot, t)$, at each t.

In this case, in a neighborhood of a point where a delta-shock forms spontaneously, one can find $y_{i\pm}(t)$ from (4.11) in the form

$$\frac{1}{y_{i+}(t) - y_{i-}(t)} \int_{y_{i-}(t)}^{y_{i+}(t)} \underline{X}(y,t)dy = x_i(t) \tag{4.13}$$

together with (3.33) and (3.26), without relying on conditions such as (3.38), (3.39). Indeed, (4.13) applies for all t.

DEFINITION. A delta-shock satisfying (4.11) is stable if there is no point $y_0 \in (y_{i-}(t),\ y_{i+}(t))$ such that

$$\underline{X}(y_0, t) = \underline{X}(y_{i\pm}(t), t) \tag{4.14}$$

and

$$\int_{y_{i-}(t)}^{y_0} (\underline{X}(y,t) - \underline{X}(y_{i\pm}(t),t))dy < \int_{y_0}^{y_{i+}(t)} (\underline{X}(y,t) - \underline{X}(y_{i\pm}(t),t))dy. \tag{4.15}$$

Such a stable delta-shock can always be continued in time, as $\underline{X}(\cdot, t)$ cannot attain or local minimum at $y_{i+}(t)$ or a local maximum at $y_{i-}(t)$. If, for example, $\underline{X}(\cdot, t)$ has a local minimum at $y_{i+}(t)$, choose y_0 the next point within $(y_{i-}(t), y_{i+}(t))$ where (4.14) holds. Then $\underline{X}(y,t) > \underline{X}(y_{i+}(t), t)$ for $y \in (y_0, y_{i+}(t))$, so the right side of (4.15) is positive. Then by (4.11), the left side of (4.11) is negative, so (4.15) holds and the delta-shock is not stable.

Thus to complete the proof of theorem 4.1 it suffices to show that stable delta-shocks can be continued satisfying the stability condition. This requires that stable delta-shocks be allowed to "fission" under certain conditions. Such fissioning may not be necessary - it is here that uniqueness of the delta-shock solution is lost, given V, W satisfying (3.1), (3.2).

Specifically, suppose that a delta-shock is stable up to some t_0 but not thereafter. Then there is a $y_0(t)$ satisfying (4.14), such that (4.15) fails for $t \leq t_0$ but holds for $t > t_0$. Then $\underline{X}(\cdot, t_0)$ is nondecreasing in a neighborhood of $y_0(t_0)$ and does not achieve a local extremum at $y_0(t_0)$. (Otherwise (4.14), (4.15) would hold at an earlier time for some other choice of y_0.)

The interval I_i terminates at t_0, and two intervals I_j, I_k begin at t_0. Subsequent references to $y_{i\pm}(t)$ or $x_i(t)$ for $t > t_0$ are only for use in the present argument. For $t - t_0 > 0$ and sufficiently small, the values of $x_i(t)$ and $y_{i\pm}(t)$ can still be determined from (4.6), (3.32) and (3.33).

The fission at t_0 is accomplished by taking initial values

$$y_{j-}(t_0) = y_{i-}(t_0),\ y_{j+}(t_0) = y_0(t_0) \tag{4.16}$$

$$y_{k-}(t_0) = y_0(t_0),\ y_{k+}(t_0) = y_{i+}(t_0) \tag{4.17}$$

$$M_j(t_0) = \int_{y_{i-}(t_0)}^{y_0(t_0)} W(y,t)dy,\ M_k(t_0) = \int_{y_0(t)}^{y_{i+}(t_0)} W(y,t)dy. \tag{4.18}$$

Since by continuity in time, using (4.11) at t_0,

$$\int_{y_{i-}(t_0)}^{y_0(t_0)} (\underline{X}(y,t) - X(y,t))dy = \int_{y_0(t_0)}^{y_{i+}(t_0)} (\underline{X}(y,t) - X(y,t))dy \tag{4.19}$$

it follows that (4.11) holds initially for the "fission fragments" M_j and M_k. Thus they are initially stable allowing for the possibility of fission into more than two fragments.

It remains to show only that the two fragments M_j and M_k will actually separate, i.e.

$$x_j(t) < x_k(t) \tag{4.20}$$

for $t - t_0$ sufficiently small and positive. For such t, (4.15) holds,

$$\int_{y_{i-}(t)}^{y_0(t)} (\underline{X}(y,t) - x_i(t)dy < \int_{y_0(t)}^{y_{i+}(t)} (\underline{X}(y,t) - x_i(t))dy, \tag{4.21}$$

so as (4.11) also holds, we find

$$\frac{1}{y_0(t) - y_{i-}(t)} \int_{y_{i-}(t)}^{y_0(t)} \underline{X}(y,t)dy < x_i(t) < \frac{1}{y_{i+}(t) - y_0(t)} \int_{y_0(t)}^{y_{0+}(t)} \underline{X}(y,t)dy. \tag{4.22}$$

Now (4.13) applies to each fragment,

$$\frac{1}{y_{j+}(t) - y_{j-}(t)} \int_{y_{j-}(t)}^{y_{j+}(t)} \underline{X}(y,t)dy = x_j(t) \tag{4.23}$$

$$\frac{1}{y_{k+}(t) - y_{k-}(t)} \int_{y_{k-}(t)}^{y_{k+}(t)} \underline{X}(y,t)dy = x_k(t). \tag{4.24}$$

By continuity, $\underline{X}(\cdot,t)$ is nondecreasing in a neighborhood of $y_0(t)$, so (4.22), (4.23), (4.23) can hold only with

$$y_{j+}(t) < y_0(t) < y_{k-}(t) \tag{4.25}$$

and (4.20) holding. Thus the proof is complete. □

Such a fission is illustrated in fig. 4.1. An entropy condition on distribution solutions, familiar in the case of zero pressure gas dynamics [B, BG, Lf, LW], is an alternative method of guaranteeing the continuation of a delta-shock solution.

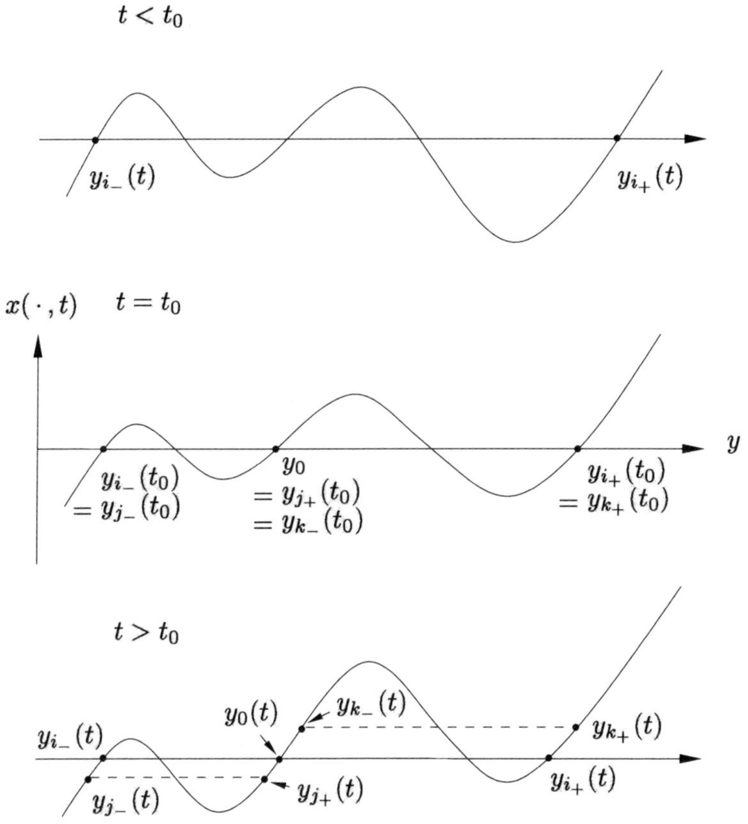

Figure 2.4.1: Fission of a delta-shock

THEOREM 4.2. *Given (V, W) satisfying (3.1), (3.2), assume that a corresponding delta-shock solution $(X, \hat{w})$ constructed from (3.23-3.33) satisfies an entropy condition on each of the delta-shocks appearing,*

$$u(W(y_{i-}(t) - 0, t)) > x_{i,t}(t) > u(W(y_{i+}(t) + 0, t)) \tag{4.26}$$

for almost all $t \in I_i$, $t \leq t_0$. Assume in addition that no delta-shocks form spontaneously at $t = t_0$. Then the delta-shock solution can be contained to some finite time $t_1 > t_0$.

REMARKS. From (3.35) and (3.33), it follows that for such a solution the y_{i+} (respectively y_{i-}) are monotone increasing (respectively decreasing) in time. Thus such solutions satisfy the "sticky particle condition"

$$X(y', t) = X(y, t), \; t > \underline{t}, \quad \text{for all } y', y, \underline{t} \tag{4.27}$$

such that $X(y', \underline{t}) = X(y, \underline{t})$.

Such a delta-shock solution is unique at least for the given weak solution (V, W) of (3.1), (3.2) [B, ERS].

PROOF. The danger to possible continuation of the delta-shock solution is that $\underline{X}(\cdot, t_0)$ attains a local minimum (respectively a local maximum) at some $y_{i+}(t_0)$ (respectively $y_{i-}(t_0)$). But given (4.23), should this happen, for $t - t_0$ sufficiently small and positive we shall have

$$x_i(t) > \underline{X}(y_{i+}(t_0, t), t) \quad (\text{respectively } x_i(t) < \underline{X}(y_{i-}(t_0), t)) \tag{4.28}$$

assuring that (4.7) can be solved for $y_{i\pm}(t)$, also satisfying (4.8). The $y_{i\pm}(t)$ determined from (4.7) will not be unique; taking y_{i+} (respectively y_{i-}) the largest (respectively smallest) such point assures that (4.8) holds. □

A sufficient condition that such an "entropy delta-shock solution" exists is the following.

THEOREM 4.3. *Assume that all of the characteristic families of the system (2.1) are linearly degenerate, and that for all* $w, w' \in D$

$$sgn(u_w(w)(w'-w)) = sgn(u(w') - u(w)) \tag{4.29}$$

and that $b(w) = b(u(w))$, *satisfies*

$$u_w(w)(b(w') - b(w)) \geq 0. \tag{4.30}$$

Assume that the initial date satisfies

$$|u(w(\cdot,0))_x| \leq c, \;\; x \in \mathbb{R} \tag{4.31}$$

and

$$u(w(\cdot,0))_{xxx} \Big|_{x=x_0} > 0 \tag{4.32}$$

for all x_0 *such that*

$$u(w(\cdot,0))_x \Big|_{x=x_0} < 0 \tag{4.33}$$

and

$$u(w(\cdot,0))_{xx} \Big|_{x=x_0} = 0. \tag{4.34}$$

Then given a weak solution (V, W) *satisfying (3.1), (3.2), there exists an entropy delta-shock solution* $(X, \hat{w})$ *continuing indefinitely in time.*

REMARKS. A simple example of a system satisfying (4.29), (4.30) is the pair of equations

$$\begin{aligned} &\rho_t + (\rho u + f_u(u))_x = 0 \\ &(\rho u)_t + (\rho u^2 + u f_u(u) - f(u))_x = 0 \end{aligned} \tag{4.35}$$

with $f_u(u) \geq 0$.

One could assume u satisfying (4.10), i.e. a linear function of w, instead of assuming linearly degenerate characteristic families.

PROOF. The condition (4.30) implies (1.45). Taking the product of (3.2) with $u_w(W)$ and using (1.45), it follows that $u(W)$ is an "entropy density" for the system (3.1), (3.2) with corresponding entropy flux zero.

Since the characteristic families of (2.1) are all linearly degenerate, so are those of (3.1), (3.2), and weak solutions of (3.1), (3.2) satisfy

$$u(W)_t = 0, \quad u(W(y,t)) = u(W(y,0)) \tag{4.36}$$

pointwise with respect to y. Then from (3.1), (4.2), (4.36)

$$V(y,t) = V(y,0) + tu(w(\cdot,0))_x \Big|_{x=\underline{X}(y,0)} V(y,0). \tag{4.37}$$

The points (y_1, t_1) at which delta-shocks may form spontaneously satisfy (3.36), (3.37). Differentiating (4.37) twice with respect to y and once with respect to t, such points (y_1, t_1) correspond to $x_0 = \underline{X}(y_1, 0)$ for which (4.33) and (4.34) hold. Then (4.32) implies that (3.38) holds and (4.33) implies (3.39), so (3.45) holds and from (3.35), delta-shocks forming spontaneously will satisfy the entropy condition (4.26) for some short time after their formation.

This reduces the proof of the theorem to that of three lemmas. Lemmas 4.4 and 4.5 imply that the entropy condition (4.26) remains satisfied in the absence of collisions of delta-shocks, while lemma 4.6 implies that the entropy condition holds immediately after each collision. Then by appeal to theorem 4.2, the entropy delta-shock solution continues indefinitely in time. Below we employ the abbreviations

$$w_\pm = w(X(y_{i\pm}(t) \pm 0, t)), u_\pm = u(w_\pm), M = M_i(t). \tag{4.38}$$

□

LEMMA 4.4. *For $u(M) = u_+ + 0$ (respectively $u_- - 0$),*

$$\frac{d}{dt}(u(M)) > 0 \quad (\textit{respectively } < 0). \tag{4.39}$$

PROOF. From $(3.59)_1$ using (4.38) and (2.1)

$$\frac{dM}{dt} = (u(M) - u_+)w_+ - (u(M) - u_-)w_- - b(w_+) + b(w_-) \tag{4.40}$$

so

$$\begin{aligned} \frac{d}{dt}u(M) = (u(M) - u_+)u_w(M)(w_+ - M) - (u(M) - u_-)u_w(M)(w_- \\ - M) - u_w(M)(b(w_+) - b(M)) + u_w(M)(b(w_-) - b(M)). \end{aligned} \tag{4.41}$$

We observe that from (4.36)

$$\begin{aligned} x_i(t) &= X(y_{i-}(t), 0) + tu_- \\ &= X(y_{i+}(t), 0) + tu_+ \end{aligned} \tag{4.42}$$

and as $X(y_{i-}(t), 0) < X(y_{i+}(t), 0)$, we have necessarily

$$u_- > u_+. \tag{4.43}$$

Consider the case $u(M) = u_+ + 0$, the other case being entirely similar. Then using (4.30) and the assumption that b depends only on $u(w)$, (4.41) becomes

$$\begin{aligned} \frac{d}{dt}u(M) &= (u_- - u_+)u_w(M)(w_- - M) + u_w(M)(b(w_-) - b(M)) \\ &\geq (u_- - u_+)u_w(M)(w_- - M). \end{aligned} \tag{4.44}$$

Now from (4.29)

$$\begin{aligned} sgn(u_w(M)(w_- - M)) &= sgn(u_- - u(M)) \\ &= sgn(u_- - u_+) \end{aligned} \tag{4.45}$$

so (4.39) follows from (4.43) (4.44) and (4.45). □

LEMMA 4.5. *For $u(M) = u_+ + 0$ (respectively $u_- - 0$)*

$$\frac{du_+}{dt} \leq 0 (\textit{respectively } \frac{du_-}{dt} \geq 0) \tag{4.46}$$

PROOF. Again we consider the first case, as the two cases are entirely similar. Recalling (3.33), the (distribution) time derivative of the second line of (4.42) is

$$u(M) = u_+ + t\frac{du_+}{dt} + \frac{d}{dt}X(y_{i+}(t), 0). \tag{4.47}$$

Recalling (3.35), that y_{i+} is a nondecreasing function of t, for $u(M) = u_+ + 0$ in (4.47) it follows that du_+/dt cannot be positive. □

Finally we consider collisions of delta-shocks. The exceptional case of the collision of more than two shocks may be regarded as an immediate succession of collisions each of two shocks, so it suffices to consider the case of the collision of two shocks.

LEMMA 4.6. *Suppose a delta-shock of singular mass M_i collides with one of singular mass M_j, producing a delta-shock initially of singular mass $M_i + M_j$. Then*

$$u(M_i + M_j) \in [u(M_i), u(M_j)]. \tag{4.48}$$

REMARKS. Thus if M_i, M_j each satisfied the entropy condition (4.26) immediately before the collision, the outgoing shock will satisfy the entropy condition immediately thereafter.

PROOF. Denote by

$$\mathcal{U}(\eta) = u((1-\eta)M_i + \eta M_j) \quad 0 \le \eta \le 1. \tag{4.49}$$

If for some $\eta_0 \in (0,1)$

$$\mathcal{U}(\eta_0) \notin [u(M_i), u(M_j)] \tag{4.50}$$

then there exists $\eta_1 \in (0,1)$ such that

$$\begin{aligned} \mathcal{U}_\eta(\eta_1) &= u_w(1-\eta_1)M_i + \eta_1 M_j)(M_j - M_i) \\ &= 0 \end{aligned} \tag{4.51}$$

and

$$\mathcal{U}(\eta_1) \notin [u(M_i)_1 u(M_j)]. \tag{4.52}$$

Abbreviating $\bar{M} = (1-\eta_1)M_i + \eta_1 M_j$ we note

$$M_j - M_i = \frac{\bar{M} - M_i}{\eta_1} = \frac{M_j - \bar{M}}{1 - \eta_1} \tag{4.53}$$

so (4.51) becomes

$$u_w(\bar{M})(\bar{M} - M_i) = u_w(\bar{M})(M_j - \bar{M}) = 0 \tag{4.54}$$

From (4.29), this implies

$$u(\bar{M}) = u(M_i) = u(M_j) \tag{4.55}$$

contradicting (4.52).

By homogeneity of u,

$$\begin{aligned} u(M_i + M_j) &= u((M_i + M_j)/2) \\ &= \mathcal{U}(1/2), \end{aligned} \tag{4.56}$$

which is within $[u(M_i), u(M_j)]$. □

Finally, we observe that in some cases the extension of the system (3.1), (3.2) to negative V is unimportant.

THEOREM 4.7. *Assume a system of the form (2.1) satisfying (1.45), and with* $b_0(w)$ *a function only of* $u(w)$. *Assume in addition that all of the associated characteristic families are linearly degenerate. Then delta-shock solutions constructed from weak solutions of (3.1), (3.2) are independent of the extension of* b *to negative* V.

PROOF. As in the proof of Theorem 4.3, the assumptions of (1.45) and linearly degenerate characteristic families imply that (4.36) and (4.37) hold. Thus the region

$$N = \{(y,t)|V(y,t) < 0\} \tag{4.57}$$

is independent of the extension of b, and the complement of N, denoted by $\bar{N}$, is simply connected. Thus for $(y,t) \in \partial N$,

$$V(y,t) = 0, \tag{4.58}$$

$$u(W(y,t)) = u(W(y,0)) \tag{4.59}$$

and from the assumption on b_0

$$b(W(y,t)) = b_0(u(W(y,0))) \tag{4.60}$$

all of which are independent of the extension of b.

Thus the solution(s) of (3.1), (3.2) within $\bar{N}$ are independent of the extension of b. Indeed it suffices to show that the singular masses $M_i(t)$ are independent of the extension of b, as the velocities of the delta-shocks are obtained from (3.33) and the $y_{i\pm}(t)$, which necessarily lie in $\bar{N}$, are determined by (3.30) or (3.35).

The points at which delta-shocks form spontaneously lie in ∂N. Differentiating (3.32) with respect to time and using (3.2), (3.35), we find

$$\begin{aligned}
\frac{dM_i(t)}{dt} &= W(y_{i+}(t)+0,t)y_{i+,t}(t) - W(y_{i-}(t)-0,t)y_{i-,t}(t) \\
&\quad + \int\limits_{y_{i-}(t)}^{y_{i+}(t)} W_t(y,t)dt \\
&= W(y_{i+}(t)+0,t)(u(M_i(t)) - u(W(y_{i+}(t)+0,t)))/V(y_{i+}(t)+0,t)) \\
&\quad + W(y_{i-}(t)-0,t)(u(W(y_{i-}(t)-0,t)) - u(M_i(t)))/V(y_{i-}(t)-0,t)) \\
&\quad + b(W(y_{i+}(t)+0,t))/V(y_{i+}(t)+0,t) \\
&\quad - b(W(y_{i-}(t)-0,t))/V(y_{i-}(t)-0,t).
\end{aligned} \tag{4.61}$$

The right side of (4.61) depends only on $M_i(t)$ and on V, W within $\bar{N}$, so $dM_i(t)/dt$ is independent of the extension of b.

At points where delta shocks collide, the singular masses simply add, so the $M_i(t)$ continue independently of the extension of b. □

5. Nonhyperbolic systems with delta-shock solutions

The remainder of this chapter is devoted to examples of systems admitting delta-shock solutions. Delta-shocks are commonly associated with systems of conservation laws with exceptional properties: a common characteristic speed, a deficiency of one eigenvector, and linearly degenerate characteristic families. The example $(1.16)_1$ shows that these conditions are not essential. Nevertheless, they do follow in an important special case, to which this section is devoted. The following theorem is proved in [S4], and the proof is not repeated here.

THEOREM 5.1. *For a system of conservation laws* $(1.1)_1$, *assume the existence of phase space coordinates* $\eta > 0, \xi \in \Omega_\xi \subset \mathbb{R}^{n-1}$ *open and bounded, such that* $w(\eta, \xi)$ *is a smooth function with the following properties:*

$$w(\eta, \xi) \text{ uniquely determines } \eta, \xi; \tag{5.1}$$

$$w \text{ and } q(w) \text{ are affine in } \eta; \tag{5.2}$$

$$\eta w_\eta(\xi) \text{ uniquely determines } \eta, \xi; \tag{5.3}$$

$$w_\eta \text{ and the } n-1 \text{ columns of } w_{\eta\xi} \text{ are everywhere linearly independent.} \tag{5.4}$$

Assume in addition that for given smooth data $w(\cdot, 0)$ *there exists a unique smooth solution of* $(1.1)_1$, *continuously dependent on the data, for some finite time which may depend on* $w(\cdot, 0)$.

Then the system $(1.1)_1$ *admits a characteristic speed* λ *of multiplicity* n, *depending only on* ξ. *At any point where* $|\lambda_\eta| \neq 0$, *there are exactly* $n-1$ *linearly independent right eigenvectors, and all of the characteristic families are linearly degenerate.*

There are representation theorems for such systems [S4,Z]; the proof of the following theorem will also not be repeated here.

THEOREM 5.2. *Assume a system* $(1.1)_1$ *with a characteristic speed* $\lambda(w)$ *of multiplicity* n, $n-1$ *linearly dependent right eigenvectors, linearly degenerate characteristic families, and such that* $\lambda_w(w)$ *never vanishes. Then there exist coordinates for phase space* $\eta > 0$, $\xi \in \Omega_\xi$ *such that* λ *depends only on* ξ, *(5.1), (5.3), (5.4) hold, and smooth functions* $f : \omega_\xi \mapsto \mathbb{R}^n$, $g : \lambda(\Omega_\xi) \mapsto \mathbb{R}^n$ *such that*

$$w(\eta, \xi) = \eta f(\xi) + g_\lambda(\lambda(\xi)), \tag{5.5}$$

$$q(w(\eta, \xi)) = \eta\lambda(\xi) f(\xi) + \lambda(\xi) g_\lambda(\lambda(\xi)) - g(\lambda(\xi)). \tag{5.6}$$

Inserting (5.5), (5.6) into $(1.1)_1$, we consider here systems of the form

$$(\eta f + g_\lambda)_t + (\eta\lambda f + \lambda g_\lambda - g)_x = 0 \tag{5.7}$$

with $f(\xi), \lambda(\xi), g(\lambda)$ given smooth functions. Zero pressure gas dynamics $(3.1)_1$ corresponds to

$$\lambda(\xi) = \xi, \; f(\xi) = \begin{pmatrix} 1 \\ \xi \end{pmatrix}, \; g(\lambda) = 0. \tag{5.8}$$

Finding delta-shock solutions for systems of the form (5.7), at least for some finite time after singularities formed spontaneously is not difficult.

LEMMA 5.3. *Under the conditions of theorem 5.2, any smooth function* $U(\lambda)$ *is an entropy density for systems (5.7), with corresponding entropy flux*

$$F(\lambda) = \lambda U(\lambda) - \int^\lambda U(\lambda')d\lambda'. \tag{5.9}$$

PROOF. As λ_w never vanishes and is perpendicular to each of the right eigenvectors, it follows that λ_w is the left eigenvector of q_w corresponding to the nondiagonal Jordan block. Thus

$$\lambda_w(w)q_w(w) = \lambda(w)q_w(w)$$

so for smooth solutions w of $(1.1)_1$,

$$\begin{aligned} U(\lambda)_t &= U_\lambda(\lambda)\lambda_w(w)w_t \\ &= -U_\lambda(\lambda)\lambda_w(w)q(w)_x \\ &= -U_\lambda(\lambda)\lambda_w(w)q_w(w)w_x \\ &= -U_\lambda(\lambda)\lambda(w)\lambda_w(w)w_x \\ &= -U_\lambda(\lambda)\lambda\lambda_x \\ &= -F(\lambda)_x \end{aligned} \tag{5.10}$$

using (5.9) in the last step. □

As all of the characteristic families of the system (5.7) are linearly degenerate, (5.10) holds as well for ordinary weak solutions of (5.7), but not for distribution solutions.

Choosing U successively equal to each component of g_λ, we find by application of lemma 5.3 that smooth solutions, or weak solutions, of (5.7) are equivalent to those of a simpler system,

$$(\eta f(\xi))_t + (\eta\lambda(\xi)f(\xi))_x = 0. \tag{5.11}$$

Identifying $u(w)$ with $\lambda(\xi)$, both (5.7) and (5.11) are of the form (2.1), except that $u(w)$, while a smooth bounded function of w, is in general not homogeneous of degree zero. This does not prevent one from affixing an additional equation of the form (3.3) in each case and switching to a Lagrangian space coordinate, obtaining systems analogous to (3.1) and (3.2).

In the case of system (5.11), the system thus obtained takes a particularly simple form

$$V_t - \lambda(\xi)_y = 0 \tag{5.12}$$

$$(\gamma f(\xi))_t = 0, \tag{5.13}$$

where

$$\gamma = \eta V \text{ wherever } V > 0. \tag{5.14}$$

A solution of (5.12), (5.13) is easily obtained. We extend the right side of (5.13) as zero also for $V \leq 0$, as theorem 4.7 does not apply. Then using (5.3) and (5.13), we have for all $y \in \mathbb{R},\ t > 0$

$$\gamma(y,t) = \gamma(y,0) \tag{5.15}$$

$$\xi(y,t) = \xi(y,0) \tag{5.16}$$

so

$$\lambda(\xi(y,t)) = \lambda(\xi(y,0)) \tag{5.17}$$

and from (5.12) and (5.17)

$$V(y,t) = V(y,0) + t\lambda(\xi(y,0))_y. \tag{5.18}$$

The initial data for V is chosen so that $\gamma(\cdot,0)=1$; then from (5.15), $\gamma=1$ everywhere.

In the case of the system (5.7), from (3.5) and (5.5), we have

$$W(y,t)=\gamma(y,t)f(\xi(y,t))+V(y,t)g_\lambda(\lambda(y,t)) \tag{5.19}$$

and (3.2) assumes the form

$$W_t-g(\lambda(\xi(y,t)))_y=0 \tag{5.20}$$

at least for $V>0$.

As the systems (5.7) and (5.11) are equivalent, for smooth or weak solutions, so are the systems (5.12), (5.13) and (5.12), (5.20) for $V\geq 0$. Having extended (5.13) to $V\leq 0$ with the right-hand side unchanged, the equivalence with (5.20) is preserved by also extending (5.20) to $V\leq 0$ without change. Indeed, it will be observed that with $\xi,\lambda(\xi)$, V given by (5.16), (5.17), (5.18), and $\gamma=1$, $W(y,t)$ obtained from (5.19) satisfies (5.20) irrespectively of the sign of $V(y,t)$.

The values of $V(y,t)$, obtained from (5.18), presumably include zero and negative values for some positive t; the procedure of section 3 is applied to recover a delta-shock solution of (5.7). A slight modification is required as λ is in general not homogeneous if degree zero is w.

The singular mass of a shock will be determined from (3.32), as previously, with $W(y,t)$ now given in (5.19). Now from (3.31) and (5.19), for $y_{i-}(t)<y<y_{i+}(t)$

$$\begin{aligned}\hat{w}(y,t)&=\tfrac{M_i(t)}{y_{i+}(t)-y_{i-}(t)}\\ &=\hat{\gamma}_i(t)f(\hat{\xi}_i(t))\end{aligned} \tag{5.21}$$

as within such an interval $X(\cdot,t),\gamma(\cdot,t),\xi(\cdot,t)$ are all independent of y. The values of $\hat{\gamma}(t),\hat{\xi}_i(t)$ are uniquely determined from (5.21), using (5.3); although the solution of the Lagrangian system (5.12), (5.2) gave $\gamma=1$ everywhere, in general the values of $\hat{\gamma}_i(t)$ will in general not be equal to one.

Finally (3.33) is modified to

$$x_{i,t}(t)=\lambda(\hat{\xi}_i(t)) \tag{5.22}$$

so that

$$X_t(y,t)=\lambda(\xi(y,t)) \tag{5.23}$$

will hold almost everywhere in $\mathbb{R}\times\mathbb{R}_+$, analogously with the definition in a distribution solution (2.4).

In this generality, there is no result that the delta-shock solution can be continued indefinitely in time. Both theorems 4.1 and 4.3 hold for zero pressure gas dynamics; this case is exceptional.

There is also no uniqueness result in this generality. A different extension of (5.20) for nonpositive V would lead, in general, to an alternative delta-shock solution. In addition there is the possibility of fissioning of delta-shocks as discussed in section 4.

6. Strictly hyperbolic, linearly degenerate pairs

The assumption (5.2) of theorem 5.1 readily allows the determination of systems with less restrictive structural conditions admitting delta-shock solutions. In particular, such systems are obtained as perturbations for $|w|$ large of the systems described by theorem 5.1.

Perhaps the simplest systems so obtained are strictly hyperbolic pairs of conservation laws with both characteristic families linearly degenerate. The system $(1.16)_1$ is a prototype of this class of systems.

The construction of delta-shock solutions is facilitated by writing a given system in the form (2.1). In this context, we present an alternative representation of such systems to that given in $(1.15)_1$.

THEOREM 6.1. *Assume a pair of conservation laws* $(1.1)_1$ *which is strictly hyperbolic with the characteristic speeds* $\lambda_\pm(w)$ *suitable coordinates for phase space. With* u, v *given by* $(1.11)_1$, *the system is of the form (2.1) with*

$$b_u = -W_v, \; b_v = -W_u \tag{6.1}$$

$$W = vw. \tag{6.2}$$

PROOF. Inserting $q = uw + b$ into $(1.14)_1$ we obtain

$$b_u = -w - vw_v, \; b_v = -vw_u; \tag{6.3}$$

then defining W by (6.2), (6.1) follows. □

From (6.1), both b and W satisfy the wave equation,

$$W_{uu} - W_{vv} = b_{uu} - b_{vv} = 0. \tag{6.4}$$

One constructs such systems by solving (6.4) for each component of W (or b) then finding the corresponding b (or W) from (6.1) and w from (6.2). The system $(1.16)_1$ corresponds to perhaps the simplest such solution, corresponding to

$$W = \begin{pmatrix} 1 \\ u \end{pmatrix}, \; b = \begin{pmatrix} 0 \\ -v \end{pmatrix} \tag{6.5}$$

The two components of w must uniquely determine u, v. For example, for the system $(1.16)_1$, $v = 1/w_1$ and $u = w_2/w_1$, so $u(w)$ is homogeneous of degree zero in w.

However, this is exceptional. We begin by modifying the definition of a distribution solution of (2.1), replacing (2.4) by

$$X_t(y,t) = \begin{cases} u(\hat{w}(y,t)/X_y(y,t)), & X_y(y,t) > 0 \\ \lim_{\epsilon\downarrow 0} u(\hat{w}(y,t)/\epsilon), & X_y(y,t) = 0 \end{cases} \tag{6.6}$$

which is equivalent to (2.4) for u homogeneous of degree zero.

The region D is chosen such that the limit in (6.6) exists, i.e., such that for all $w \in D$, $w/\epsilon \in D$ for all sufficiently small positive ϵ, depending on w, and the limit in (6.6) is defined.

This restriction precludes the obtaining of delta-shock solutions for some systems obtained in this manner, for example,

$$\begin{cases} (\frac{1}{v})_t + (\frac{u}{v})_x = 0 \\ u_t + (\frac{u^2-v^2}{2})_x = 0 \end{cases} \tag{6.7}$$

for which $u(w) = w_2$ and the limit in (6.6) is not defined for all $w_2 \neq 0$.

There are systems of this class other than $(1.16)_1$ which do admit delta-shock solutions. An example is

$$\begin{cases} (\frac{1}{v})_t + (\frac{u}{v})_x = 0 \\ (\frac{u^3}{v} + 3uv)_t + (\frac{u^4}{v} - v^3)_x = 0. \end{cases} \tag{6.8}$$

As in the preceding section, the construction of delta-shock solutions is greatly facilitated by the linearly degenerate characteristic families, as smooth or weak solutions of any pair of conservation laws in this class are equivalent to those of $(1.16)_1$, and this will remain so under a change to a Lagrangian space variable. For the system $(1.16)_1$, the corresponding form of (3.3), (3.4) is

$$\begin{cases} \rho_t + (\rho u)_x = 0 \\ (\frac{1}{v})_t + (\frac{u}{v})_x = 0 \\ (\frac{u}{v})_t + (\frac{u^2}{v} - v)_x = 0 \end{cases} \tag{6.9}$$

and the corresponding form of (3.1), (3.2) is

$$\begin{cases} V_t - u_y = 0 \\ (\frac{V}{v})_t = 0 \\ (V\frac{u}{v})_t - v_y = 0 \end{cases} \tag{6.10}$$

valid for $V > 0$.

The system $(1.16)_1$ does not satisfy (1.45), so theorem 4.7 cannot be applied; the extension of (6.10) to negative V will affect the delta-shock solutions ultimately obtained. In this context we appeal to theorem 3.2. The specific choice $(1.25)_1$ for an entropy density $U(w)$ indeed satisfies $U_w(w)$ homogeneous of degree zero, and with $b_1(w) = \binom{0}{-1/w_1}$ obtained from (6.5), using (3.5), the left side of (3.1) is evaluated as $\binom{0}{-W_1}$. However, we may choose initial data $V(y,0) = v(y,0)$ so that

$$W_1(y,0) = \frac{V(y,0)}{v(y,0)} = 1 \tag{6.11}$$

then $W_1(y,t) = 1$ as long as $V(y,\cdot)$ remains positive from (6.10). So the condition (3.10) is satisfied with $b^* = \binom{0}{-1}$, and we obtain from theorem 3.2 an extension of (6.10)

$$\begin{cases} V_t - u_y = 0 \\ u_t - V_y = 0 \\ V = v \end{cases} \tag{6.12}$$

for all V. An alternative, selected for simplicity, is

$$\begin{cases} V_t - u_y = 0 \\ u_t = \begin{cases} V_y, & V > 0 \\ 0, & V \le 0 \end{cases} \\ V = v \end{cases} . \tag{6.13}$$

Denote by V, v, u a solution of (6.12) or (6.13) for given initial data. We may apply the procedure of section 3 to recover a delta-shock solution, for at least some finite time after delta-shocks form spontaneously.

From (3.5) we have

$$W(y,t) = V(y,t)w(u,(y,t), v(y,t)), \tag{6.14}$$

i.e.,

$$W(y,t) = \begin{pmatrix} 1 \\ u(y,t) \end{pmatrix} \tag{6.15}$$

for the system $(1.16)_1$, whereas

$$W(y,t) = \begin{pmatrix} 1 \\ u^3(y,t) + 3u(y,t)v^2(y,t) \end{pmatrix} \tag{6.16}$$

for the system (6.8). For a delta-shock at location $x_i(t)$, the values of $y_{i\pm}(t), M_i(t)$, $X(\cdot,t), w\hat{}(\cdot,t)$ are determined by (3.24-3.32). However as u is in general not homogeneous of degree zero in w, W we find $x_{i,t}(t)$ from (6.6) with $X_y = 0$, as is the case for $y \in (y_i(t), y_{it}(t))$. Thus (3.33) is replaced by

$$x_{i,t}(t) = \lim_{\epsilon \downarrow 0} u(M_i(t)/\epsilon), \tag{6.17}$$

which is $M_i^{(2)}(t)/M_i^{(1)}(t)$ for the system $(1.16)_1$ and $(M_i^{(2)}(t)/M_i^{(1)}(t))^{1/3}$ for the system (6.8). This difference illustrates the effect of exchanging conserved quantities on distribution solutions of systems of conservation laws, even these with all characteristic families linearly degenerate.

Finally we give a numerical example showing the effect of different extensions (6.12), (6.13) of the system (6.10) to negative V. In this case, the system $(1.16)_1$ satisfies (4.10), so that theorem 4.1 assures that a delta-shock solution can be continued indefinitely.

For initial data as shown in fig. 6.1., the corresponding $W^{1,\infty}$ "weak" solution of (6.12) is shown in fig. 6.2, and that of (6.13) in fig. 6.3. The regions of negative V are quite different. Indeed, by application of the above procedure, for the system (6.12) the speed of the resulting delta-shock is -1 for $t \geq 5/4$, whereas for the system (6.13) the speed of the delta-shock approaches $1 - 4\sqrt{3}/3$ for t large. We omit the computations.

We conclude that in general, a well-posed initial value problem for a system admitting delta-shock solutions must include the extension of the system (3.2) to the region $V \leq 0$, analogously with the specification of an entropy condition for weak solutions.

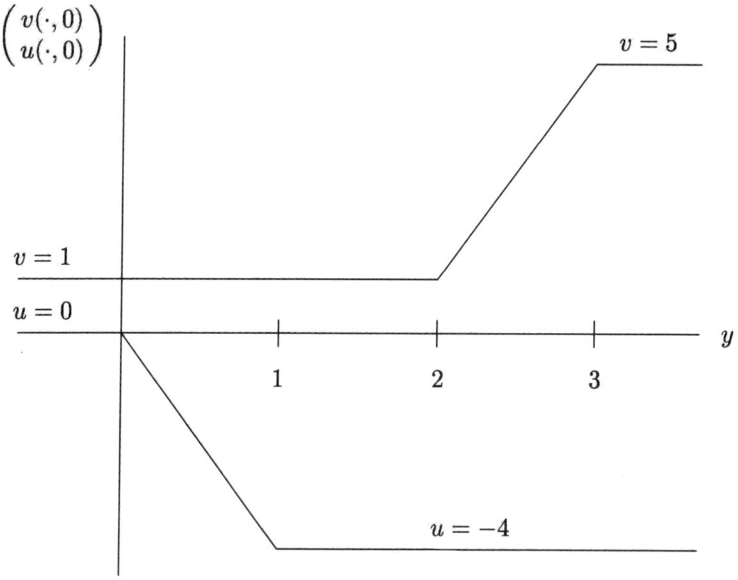

Figure 2.6.1: Initial data

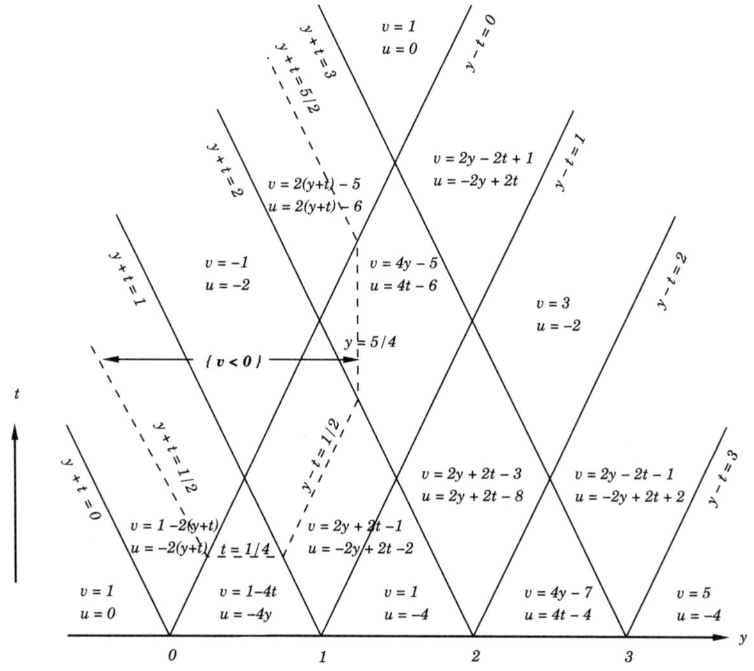

Figure 2.6.2: Solution of (6.12) with $\{v < 0\}$ region

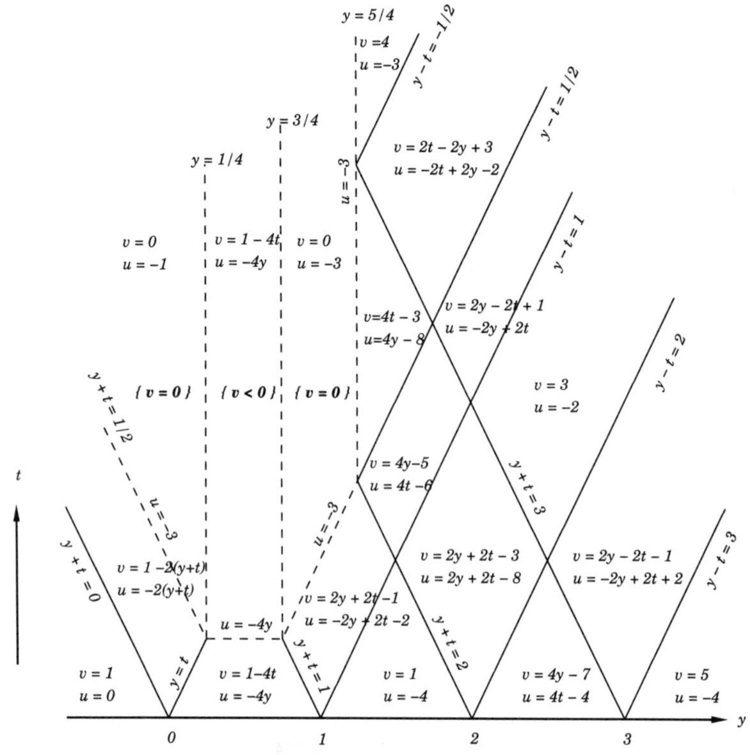

Figure 2.6.3: Solution of (6.13) with $\{v < 0\}$ and $\{v = 0\}$ regions

7. Additional examples

The previous discussion has focused on systems admitting single-valued delta-shock solutions for a wide class of initial data, and the implied limitations on the structure of such systems. Relaxation of these conditions permits the identification of delta-shock solutions with a larger class of systems of conservation laws.

Reference was made in section 1 of chapter 1 to unsolvable Riemann initial value problems in the class of weak solutions. More generally, given a smooth or piecewise smooth weak solution of a system in the form (3.1), (3.2) such that $V(\cdot, 0)$ is everywhere positive but with V assuming negative values after finite time, the procedure of section 3 will often produce a single-valued delta-shock solution of the corresponding system (3.4). While this approach obtains delta-shock solutions for systems with diverse structural conditions, the delta-shock solution is associated with special initial data and a controversial change of space coordinates.

An alternative approach, leading to the appearance of delta-shock solutions while avoiding changes of space variables, is to embed a system of the form $(1.1)_1$ in an uncoupled system

$$w_t + q(w)_x = 0, \tag{7.1}$$

$$\rho_t + (q_w(w)\rho)_x = 0. \tag{7.2}$$

The corresponding initial data is restricted to satisfy

$$\rho(\cdot, 0) = w_x(\cdot, 0). \tag{7.3}$$

Then given a weak solution w, piecewise smooth and uniformly bounded but with jump discontinuities, one identifies

$$\rho(\cdot, t) = w_x(\cdot, t) \tag{7.4}$$

in the sense of distribution derivative.

A jump discontinuity in w, on a trajectory $x_i(t)$, is identified with a delta-shock of singular mass

$$M_i(t) = \begin{pmatrix} 0 \\ w(x_i(t) + 0, t) - w(x_i(t) - 0, t) \end{pmatrix} \tag{7.5}$$

for the systems (7.1) and (7.2).

In the case $n = 1$, so that each of (7.1), (7.2) is a scalar conservation law, the system (7.1), (7.2) is within the class $(4.28)_1$, identifying $u = q_w(w)$, and the related discussion applies.

For larger n, with w, ρ vector valued, ρ, w obtained from (7.1), (7.4) still qualify as weak limits in the space of measures of approximate solutions of (7.1), (7.2). This is clear for w; we consider (7.2) in a neighborhood of a trajectory $x_i(t)$ where w has a jump discontinuity.

For $\epsilon > 0$, denote by

$$w^\epsilon(x,t) = \begin{cases} w(x,t), |x - x_i(t)| \geq \epsilon \\ (\frac{1}{2} + \frac{x_i(t)-x}{2\epsilon})w(x_i(t) - 0, t) \end{cases} \tag{7.6}$$

$$+ (\frac{1}{2} + \frac{x - x_i(t)}{2\epsilon})w(x_i(t) + 0, t), \ \ |x - x_i(t)| < \epsilon$$

$$\rho^\epsilon(x,t) = \begin{cases} \rho(x,t), & |x - x_i(t)| \geq \epsilon \\ \frac{w(x_i(t)+0,t)-w(x_i(t)-0,t)}{2\epsilon}, & |x - x_i(t)| < \epsilon. \end{cases} \tag{7.7}$$

Then for $\theta \in C^1 : \mathbb{R} \times \mathbb{R}_+ \mapsto \mathbb{R}^n$ with support in a neighborhood of a point $(x_i(t_0), t_0)$ for arbitrary fixed t_0, denote by

$$\phi(y,t) = \theta(x_i(t) + y, t). \tag{7.8}$$

Then using (7.8), then (7.6) and (7.7)

$$\begin{aligned} & \int_{\mathbb{R}_+} \int_{\mathbb{R}} (\theta_t \cdot \rho^\epsilon + \theta_x \cdot q_w(w^\epsilon)\rho^\epsilon) dx\ dt \\ = & \int_{\mathbb{R}_+} \int_{\mathbb{R}} (\phi_t \cdot \rho^\epsilon - x_{i,t}\phi_y \cdot \rho^\epsilon + \phi_y \cdot q_w(w^\epsilon)\rho^\epsilon) dy\ dt \\ = & \int_{\mathbb{R}_+} \int_{|y|>\epsilon} (\phi_t \cdot \rho - x_{i,t}\phi_y \cdot \rho + \phi_y \cdot q_w(w)\rho) dy\ dt \\ + & \int_{\mathbb{R}_+} \int_{|y|<\epsilon} (\phi_t \cdot \rho^\epsilon - x_{i,t}\phi_y \cdot \rho^\epsilon + \phi_y \cdot q_w(w^\epsilon)\rho^\epsilon) dy\ dt. \end{aligned} \tag{7.9}$$

As ρ satisfies (7.2) in the region $|y| < \epsilon$, the first integral in (7.9) is equal to

$$\begin{aligned} & \int_{\mathbb{R}_+} [\phi(-\epsilon,t) \cdot (q_w(w(x_i(t) - \epsilon, t) - x_{i,t}(t))\rho(x_i(t) - \epsilon, t) \\ - & \phi(\epsilon,t) \cdot (q_w(w(x_i(t) + 0, t)) - x_{i,t}(t))\rho(x_i(t) + \epsilon, t)]\ dt \\ = & -\int_{\mathbb{R}_+} \phi(0,t) \cdot ((q_w(w(x_i(t) + 0, t)) - x_{i,t}(t))\rho(x_i(t) + 0, t) \\ - & (q_w(w(x_i(t)) - 0, t)) - x_{i,t}(t))\rho(x_i(t) - 0, t)\ dt + o(1) \end{aligned} \tag{7.10}$$

as $\epsilon \downarrow 0$.

Now from (7.6) and (7.7), we observe that for $|x - x_i(t)| < \epsilon$, corresponding to $|y| < \epsilon$

$$\rho^\epsilon(x,t) = w_x^\epsilon(x,t) \tag{7.11}$$

$$q_w(w^\epsilon(x,t))\rho^\epsilon(x,t) = q(w^\epsilon(x,t))_x \tag{7.12}$$

$$\begin{aligned} \rho_t^\epsilon(x,t) &= \frac{1}{2\epsilon}(w_t(x_i(t) + 0, t) - w_t(x_i(t) - 0, t) \\ &+ x_{i,t}(t)(w_x(x_i(t) + 0, t) - w_x(x_i(t) - 0, t))) \end{aligned} \tag{7.13}$$

and that $\rho^\epsilon(\cdot,t), w_t^\epsilon(\cdot,t), q(w^\epsilon(\cdot,t))_x$ are all bounded L_1^{loc} functions uniformly with respect to ϵ, t.

Using (7.13), the first term in the second integral in (7.9) is

$$\begin{aligned}
\int_{\mathbb{R}_+}\int_{|y|<\epsilon} \phi_t\rho^\epsilon dy\, dt &= -\int_{\mathbb{R}_+}\int_{|y|<\epsilon} \phi\rho^\epsilon_t dy\, dt \\
&= -2\epsilon\int_{\mathbb{R}_+} \phi(0,t)\rho^\epsilon_t(\cdot,t)dt + o(1) \\
&= -\int_{\mathbb{R}_+} \phi(0,t)\cdot(w_t(x_i(t)+0,t) - w_t(x_i(t)-0,t) \\
&+ x_{i,t}(t)(w_x(x_i(t)+0,t) - w_x(x_i(t)-0,t)))\, dt \\
&+ o(1) \\
&= \int_{\mathbb{R}_+} \phi(0,t)\cdot((q_w(w(x_i(t)+0,t) - x_{i,t}(t))w_x(x_i(t)+0,t) \\
&- q_w(w(x_i(t)-0,t) - x_{i,t}(t))w_x(x_i(t)-0,t))dt + o(1) \qquad (7.14)
\end{aligned}$$

since w satisfies (7.1). Then from (7.4), the sum of (7.10) and (7.14) is $o(1)$ as $\epsilon \downarrow 0$.

The last two terms in (7.9) are

$$\begin{aligned}
&\int_{\mathbb{R}_+}\int_{|y|<\epsilon} \phi_y\cdot(q_w(w^\epsilon)\rho^\epsilon - x_{i,t}\rho^\epsilon)dy\, dt \\
&= \int_{\mathbb{R}_+} \phi_y(0,t)\int_{|y|<\epsilon} (q_w(w^\epsilon)\rho^\epsilon - x_{i,t}\rho^\epsilon dy\, dt + o(1) \\
&= \int_{\mathbb{R}_+} \phi_y(0,t)\cdot\int_{|y|<\epsilon} (q(w^\epsilon)_x - x_{i,t}w^\epsilon_x)dy\, dt + o(1) \qquad (7.15)
\end{aligned}$$

using (7.11), (7.12) in the last step. Evaluating the integrals in (7.15) we obtain, using (7.6)

$$\begin{aligned}
&\int_{\mathbb{R}_+} \phi_y(0,t)\cdot[q(w^\epsilon(x_i(t)+\epsilon-0,t)) - q(w^\epsilon(x_i(t)-\epsilon+0,t)) \\
&- x_{i,t}(t)(w^\epsilon(x_i(t)+\epsilon-0,t) - w^\epsilon(x_i(t)-\epsilon+0,t))]\, dt \\
&= \int_{\mathbb{R}+} \phi_y(0,t)\cdot[q(w(x_i(t)+0,t)) - q(w(x_i(t)-0,t)) \\
&- x_{i,t}(t)(w(x_i(t)+0,t) - w(x_i(t)-0,t))]\, dt = 0 \qquad (7.16)
\end{aligned}$$

as w, a weak solution of (7.1), satisfies the ordinary Rankine-Hugoniot relation. So $w^\epsilon, \rho^\epsilon$ obtained from (7.6), (7.7) are indeed approximate solutions of (7.1), (7.2).

A need of delta-shock solutions which are not single valued is also anticipated. In view of the paucity of available examples, the discussion here is limited to a general explanation, illustrated by an example.

Consider a system of conservation laws with real characteristic speeds but such that smooth data initial value problems cannot be solved within the class of weak solutions. Assume that the system satisfies $(4.1)_1$, that is, $|w|$ majorizes $|q(w)|$ for $|w|$ large, $w \in D$, where D is unbounded and is such that one seeks "solutions" as weak limits of approximations assuming values within D.

Piecewise smooth such solutions will be of the form $(3.20)_1$, and $(3.57)_1$, in particular with the singular part of the flux a scalar multiple of the singular part of the density. Thus the singular mass of a delta-shock associated with a single-valued delta-shock solution must be within the set $\mathcal{M}$ as described in section 1, satisfying (1.8) in particular.

The set $\mathcal{M}$ so obtained may be too small for solution of initial value problems in the class of single-valued delta-shock solutions. Since the singular masses $M_i(t)$ of delta-shocks must satisfy $(3.59)_1$ simply to conserve material, one presumably requires $\mathcal{M}$ to contain open subsets of $\mathbb{R}^n$. This can easily be incompatible with the requirement (1.8).

In such cases we propose to admit multiple-valued delta-shock solutions with singular mass $M_i(t) \in \mathcal{M}'$, where $\mathcal{M}' \subset \mathbb{R}^n \setminus \{0\}$ is the set of values M such that there exists a sequence of nonnegative Young measures $d\nu_m^\epsilon$, $\epsilon > 0$, on D with the following properties.

$$\int_D d\nu_M^\epsilon = 1 \tag{7.17}$$

$$d\nu_M^\epsilon \to 0 \text{ as } \epsilon \downarrow 0 \text{ on compact subsets of } D \tag{7.18}$$

$$\epsilon \int_D w d\nu_M^\epsilon = M \tag{7.19}$$

$$\lim_{\epsilon \downarrow 0} \epsilon \int_D q(w) d\nu_M^\epsilon = s(M)M \tag{7.20}$$

for a scalar $s(M)$, identified as the speed of propagation of the delta-shock.

The construction of such solutions follows a familiar procedure. We assume that for any $M \in \mathcal{M}'$ and $\epsilon > 0$ sufficiently small depending on M, then $M/\epsilon \in D$. The fluid velocity $u : D \to \mathbb{R}$ has to be selected to satisfy

$$\lim_{\epsilon \downarrow 0} u(M/\epsilon) = s(M), \tag{7.21}$$

then one obtains

$$b(w) = q(w) - u(w)w \tag{7.22}$$

so that the given system is nominally of the form (2.1), although $b(D)$ will not in general be bounded.

Nonetheless, if one can solve the related system (3.1), (3.2), the procedure of section 3 can be applied formally, to obtain a delta-shock solution, replacing (3.33) by

$$x_{i,t}(t) = s(M_i(t)). \tag{7.23}$$

In addition to other potential difficulties (e.g., ability to continue the $y_{i,\pm}(t)$), the ability to continue such a solution will be lost should $M_i(t)$ fail to remain within $\mathcal{M}'$.

The functions $\hat{w}, X$ obtained in this manner will satisfy (2.5) and (6.6); we omit the proof.

The measures $d\nu_M^\epsilon$ are left unspecified. In any example, the choice thereof will require justification.

As an example, consider the Euler system of fluid dynamics with an equation of state

$$E = VS \tag{7.24}$$

implying

$$P = -S, \; T = V \tag{7.25}$$

where E, V, S, P, T are the internal energy, specific volume, thermodynamic entropy, pressure and temperature. We thus obtain a system

$$\begin{aligned} \rho_t + (\rho u)_x &= 0 \\ (\rho u)_t + (\rho u^2 - S)_x &= 0 \\ (\frac{1}{2}\rho u^2 + S)_x + (\frac{1}{2}\rho u^3)_x &= 0. \end{aligned} \tag{7.26}$$

The system (7.26) admits u as a characteristic speed of multiplicity three, to which corresponds a single eigenvector and a linearly degenerate characteristic family. Thus ordinary weak solutions of (7.26) also satisfy the entropy equation

$$(\rho S)_t + (\rho u S)_x = 0. \tag{7.27}$$

Assuming that bounded values of u are obtained, the system (7.26) satisfies $(4.1)_1$, so a delta-shock solution is anticipated. Thus we consider the set $\mathcal{M}$ of possible values of the singular mass of a delta-shock associated with a single-valued delta-shock solution.

For a singular mass of the form

$$M = \begin{pmatrix} \beta_1 \\ \beta_2 \\ \beta_3 \end{pmatrix} \in \mathbb{R}^3 \tag{7.28}$$

from (7.26) we readily obtain

$$q(M) = \begin{pmatrix} \beta_2 \\ \frac{3}{2}\frac{\beta_2^2}{\beta_1} - \beta_3 \\ \frac{1}{2}\frac{\beta_2^3}{\beta_1^2} \end{pmatrix} \tag{7.29}$$

so (1.8) can hold only for

$$\mathcal{M} = \left\{ M \mid \beta_1 > 0, \beta_3 = \frac{1}{2}\frac{\beta_2^2}{\beta_1} \right\} \tag{7.30}$$

which corresponds simply to a positive singular part of ρ and zero singular part of S.

The set of $\mathcal{M}$ does not contains any open subsets of $\mathbb{R}^3$, and is indeed too small to construct single-valued delta shock solutions. In particular, for a delta-shock satisfying $(3.59)_1$, we have

$$\begin{aligned} \beta_{1,t} &= x_{i,t}(\rho_+ - \rho_-) - \rho_+ u_+ + \rho_- u_- \\ \beta_{2,t} &= x_{i,t}(\rho_+ u_+ - \rho_- u_-) - \rho_+ u_+^2 - \rho_- u_-^2 - S_+ + S_- \\ \beta_{3,t} &= x_{i,t}(\frac{1}{2}\rho_+ u_+^2 - \frac{1}{2}\rho_- u_-^2 + S_+ - S_-) - \frac{1}{2}\rho_+ u_+^3 + \frac{1}{2}\rho_- u_-^3 \end{aligned} \tag{7.31}$$

with $x_{i,t}$ the shock speed and $\rho_\pm, u_\pm, S_\pm$ the limiting values on the two sides of the shock.

For $M \in \mathcal{M}$, comparing (7.28) and (7.29), the requirement that $q(M) = x_{i,t}M$ implies

$$x_{i,t} = \beta_2/\beta_1; \tag{7.32}$$

and from (7.31), (7.32) we readily compute

$$\frac{d}{dt}\left(\beta_3 - \frac{1}{2}\frac{\beta_2^2}{\beta_1}\right) = \frac{1}{2}\rho_+(x_{i,t} - u_+)^3 + \frac{1}{2}\rho_-(u_- - x_{i,t})^3 \tag{7.33}$$

which in general is not zero. So in general $M_i(t)$ will not remain in $\mathcal{M}$, and single-valued delta-shock solutions are not possible for the system (7.26).

We observe that for a shock satisfying the expected entropy condition

$$u_- > x_{i,t} > u_+, \tag{7.34}$$

the right side of (7.33) is positive. This motivates a selection of the set $\mathcal{M}'$.

$$\mathcal{M}' = \left\{ M \mid \beta_1 > 0, \beta_3 \geq \frac{1}{2}\frac{\beta_2^2}{\beta_1} \right\}. \tag{7.35}$$

Then for $M \in \mathcal{M}'$ of the form (7.28), we can satisfy (7.17-7.20) with the choice

$$d\nu_M^\epsilon = \delta(\rho - \beta_1/\epsilon)(\frac{\delta(u - s - \gamma) + \delta(u - s + \gamma)}{2})\delta(S - \beta_1\gamma^2/\epsilon)d\rho\, du\, dS, \tag{7.36}$$

with

$$s = \beta_2/\beta_1, \tag{7.37}$$

and

$$\gamma^2 = \frac{2}{3}(\beta_3 - \frac{1}{2}\beta_2^2/\beta_1). \tag{7.38}$$

The measure (7.36) admits a simple physical interpretation. Half of the condensed material effectively has velocity $s + \gamma$, half $s - \gamma$. In addition, the singular part of S is just γ^2 times the singular part of ρ.

The parameters $s(M), \gamma(M)$ given in (7.37), (7.38) have been obtained, of course, so that the measure (7.36) satisfies (7.20).

Finally we note that for the system (7.26), the corresponding system (3.1), (3.2) is

$$\begin{aligned} V_t - u_y &= 0 \\ u_t - S_y &= 0 \\ (\frac{1}{2}u^2 + SV)_t - (uS)_y &= 0 \end{aligned} \tag{7.39}$$

which is very easily solved for given initial data, as smooth or weak solutions of (7.39) also satisfy

$$S_t = 0 \tag{7.40}$$

obtained from (7.27). Thus for a fairly broad class of initial data, we can construct a delta-shock solution of (7.26) satisfying an entropy condition (7.34), at least for some finite time after singularities form.

CHAPTER 3

Singular shocks

1. Origin and structure

Singular shocks, like delta-shocks, appear as weak limits of approximate solutions of systems $(1.1)_1$. They are anticipated of the form $(3.20)_1$ satisfying $(3.59)_1$ and $(3.60)_1$ in particular. The weak limit of the corresponding flux functions is again of the form $(3.57)_1$.

A distinction between singular shocks and delta-shocks is given and discussed in section 4 of chapter 1. In the case of singular shocks, the density approximations $w^\epsilon(\cdot,t)$ are locally bounded measures on $\mathbb{R}$ uniformly with respect to ϵ, and converge weakly in the space of measures on $\mathbb{R}$ as $\epsilon \downarrow 0$, but the corresponding flux approximations $q(w^\epsilon(\cdot,t))$ are not locally bounded measures uniformly in ϵ and converge weakly only in the sense of distributions.

This difficulty has prevented the formulation of a definition of singular shock solutions analogous to that given for delta-shocks in section 2 of chapter 2. Singular shock solutions are defined as limits of viscous structures, by a direct extension of the notion of admissible weak solutions of $(1.1)_1$.

DEFINITION. **A singular shock solution of a system $(1.1)_1$ is a measure w^0 of the form $(3.20)_1$, the limit of a sequence of approximations w^ϵ satisfying the following:**

(1.1) $w^\epsilon(\cdot,t) \in L_1^{loc}$ **uniformly with respect to ϵ, pointwise in** t;

$$w^\epsilon(\cdot,t) \rightharpoonup w^0(\cdot,t) \tag{1.2}$$

as $\epsilon \downarrow 0$ weakly in the space of measures on $\mathbb{R}$, pointwise with respect to t;

$$w_t^\epsilon(\cdot,t) + q(w^\epsilon(\cdot,t))_x - \epsilon(Aw_x^\epsilon(\cdot,t))_x \rightharpoonup 0 \tag{1.3}$$

as $\epsilon \downarrow 0$ weakly in the space of measures on $\mathbb{R}$, pointwise with respect to t, for some positive definite "viscosity matrix" A.

The form of convergence shown in (1.3) is determined by expedience, and indeed may be improved in the future. Nonetheless, this condition has resulted in the identification of a sufficient class of singular shocks for solving initial value problems for some systems (1.1) otherwise unsolvable, and has produced results remarkably consistent with numerical computations [SS].

The form of viscosity matrix A has also been left unspecified in (1.3). This is arguably fatuous, as in practice results are known only (one example, shown in section 2 below, excepted) for pairs of conservation laws, with either $A = I$ [S5, SSS] or $A = tI$ [KK2, KK3], I the identity matrix. We shall see below that the structural conditions of a given system imply strong restrictions on the possible form of A. This is hardly surprising, as finding viscous structures for strong, ordinary shocks, in general, requires additional assumptions on the viscosity matrix [Mo2]. Indeed, admissible weak solutions of $(1.1)_1$ often if not always satisfy (1.1-1.3), and, in addition, the limit w is a bounded measurable function. Thus the appearance of singular shocks, with w of the form $(3.20)_1$ with nonzero singular mass(es) M_i has been limited to systems with somewhat exceptional features. We first briefly outline the situation.

Systems of conservation laws admitting singular shock solutions typically combine two structural conditions ordinarily not associated with the same system: a strictly convex entropy density $U(w)$, in much if not all of phase space; and a Hugoniot locus $\Gamma(w_0)$, given by $(1.7)_1$, containing branches which do not continue to infinity. Often if not always, $\Gamma(w_0)$ is bounded for fixed w_0. The paradoxical nature of this combination typically appears in the phase portraits of the "viscous profile system" associated with the given system of conservation laws $(1.1)_1$. The corresponding viscous profile system is an autonomous dynamical system of the form

$$S(w_0; s) \quad Aw' = q(w) - q(w_0) - s(w - w_0), \tag{1.4}$$

primes denoting differentiation and $w_0 \in \mathbb{R}^n$, $s \in \mathbb{R}$ given. We assume the viscosity matrix A independent of x, t for simplicity, but it might depend on w.

Assuming $w \in H$ and $A = A(w)$ such that

$$U_{ww}(w)A(w) > 0, \tag{1.5}$$

the same assumption as made in [Mo2], the system (1.4) admits a (nondecreasing along trajectories) Lyapunov function

$$\mathcal{L}(w) = U_w(w)(q(w) - sw) - (F(w) - sU(w)) - U_w(w)(q(w_0) - sw_0), \tag{1.6}$$

with $F(w)$ the entropy flux corresponding to U. Equation (1.6) is readily obtained by writing (1.4) in symmetric dependent variables [Go, Mo1, Mo2]; the details are unnecessary and omitted here.

Thus all limit sets of bounded semiorbits of (1.4) within H are single critical points. Critical points of (1.4), other than w_0, correspond to ordinary shocks of speed s connecting w_0 with the other critical point.

As against this, for a pair of conservation laws

$$\begin{aligned} u_t + f(u, v)_x &= 0 \\ v_t + g(u, v)_x &= 0 \end{aligned} \tag{1.7}$$

i.e., with

$$w = \begin{pmatrix} u \\ v \end{pmatrix}, \quad q(w) = \begin{pmatrix} f(u, v) \\ g(u, v) \end{pmatrix} \tag{1.8}$$

with u, v, f, g scalar valued, the condition of bounded $\Gamma(w_0)$ precludes solutions of the Rankine-Hugoniot relations

$$\begin{aligned} s(u - u_0) &= f(u, v) - f(u_0, v_0) \\ s(v - v_0) &= g(u, v) - g(u_0, v_0) \end{aligned} \tag{1.9}$$

with $|w - w_0|$ sufficiently large, uniformly with respect to s. Assuming that $|q(w)|$ increases faster than $|w|$ for $|w|$ large, (implicit in the definition of singular shock in section 4, chapter 1), without loss of generality we infer from the boundedness of Γ

$$\lim_{|w|\to\infty} \inf (ug(u,v) - vf(u,v))/|w|^2 > 0. \tag{1.10}$$

Then using (1.8) and (1.10), with θ determined from $\tan\theta = v/u$, along trajectories determined by (1.4) we have

$$\begin{aligned} \theta' &= \frac{(-v \quad u)A^{-1}\binom{f}{g}}{|w|^2} + \text{lower order terms as } |w| \to \infty \\ &> 0 \end{aligned} \tag{1.11}$$

compatible with (1.10) for sufficiently large $|w|$, uniformly with respect to θ, if A is a positive scalar multiple of the identity matrix. Indeed, (1.11) follows from (1.10) only for this choice of A.

The condition (1.11) is uniform with respect to θ for s in bounded intervals, as only the lower order terms in (1.11), obtained from (1.4), depend on s. One examines the phase portraits of (1.4) as s varies. For $s << 0$ (resp. $s >> 0$) the boundedness of Γ implies that w_0 is the only critical point, a repeller (resp. attractor). But w_0 cannot be a global repeller for $s << 0$ and a global attractor for $s >> 0$ because then (1.11) would imply the appearance of periodic orbits at some intermediate value of s, incompatible with the existence of $\mathcal{L}(w)$ given in (1.6).

We infer that for some value of s, there are trajectories of (1.4) which approach infinity in both directions, occupying a region $\Xi(s) \subset \mathbb{R}^2$. By continuity this is true for all s. The trajectories assuming values in $\mathbb{R}^2 - \overline{\Xi(s)}$ have at least one finite limit set, necessarily w_0 as the α limit set for $s << 0$ and w_0 as the w-limit set for $s >> 0$. It follows by continuity of the phase portraits with respect to s that for some intermediate value of s, the set of complete, bounded trajectories of (1.4) is itself an unbounded set in $\mathbb{R}^2$. We omit justification of this here, as this is done in the proof of theorem 1.6 below and in [S5]. This implies that one may find $w_\pm$ in a bounded region of $\mathbb{R}^2$ such that the set of travelling wave solutions of

$$w_t + q(w)_x = \epsilon(Aw_x)_x, \; w(\pm\infty, t) = w_\pm \tag{1.12}$$

is not pointwise bounded, even for $\epsilon > 0$ fixed. This conclusion is easily obtained by setting $w(x,t) = w(x - st)$ in (1.12) and identifying $\eta = (x - st)/\epsilon$ as the independent variable in (1.4).

The above discussion obviously does not suffice to determine the structural conditions identified with singular shocks. It is intended as a guide to the more precise treatment which follows.

As in the case of delta-shocks, our treatment focuses on the set $\mathcal{M} \subset \mathbb{R}^n$ containing the possible values of singular mass M. This is not surprising, recalling the discussion at the end of section 3 of chapter 1; the structural conditions of a system of conservation laws must provide at least one equation between the limiting values at a singularity, the speed of propagation, and the singular mass.

For delta-shocks, the speed s is a function of the n-vector singular mass M, as discussed in section 1 of chapter 2. For some singular shock systems, such as the model problem $(3.2)_1$, the needed relation can be obtained by inspection, as done in section 4 of chapter 1. A subspace of $\mathbb{R}^n$ containing $\mathcal{M}$ may also be determined by the symmetry group of a given system, as discussed in section 6 of chapter 1.

More generally, for singular shock systems, the direction of the singular mass vector is determined by conditions on the entropy density. The framework discussed below describes the best known examples of singular shock systems. Additional examples are included in section 2.

The entropy density U does not have to be strictly convex throughout phase space; we assume $n = 2$ and a division of phase space

$$\mathbb{R}^2 = H \cup N$$

$$H = \{\begin{pmatrix} u \\ v \end{pmatrix} \mid v < \overline{v}(u)\}, \overline{v}(u) \in [0, \infty] \tag{1.13}$$

with U strictly convex in $w = \binom{u}{v}$ within H. In addition, we assume that for $w \in H$

$$U(w) > 0, \tag{1.14}$$

and

$$U_v(w) < 0; \tag{1.15}$$

for notational simplicity we designate

$$U \mid_N \overset{def}{=} 0. \tag{1.16}$$

The entropy density U cannot be uniformly convex in w, for this would imply L_2 bounds on the approximate solutions and preclude weak limits as distributions. The growth of U at infinity is indeed of particular importance. We assume U satisfying

$$\liminf_{v \to -\infty} \frac{v}{U(u, v)} = 0 \text{ uniformly in } u, \tag{1.17}$$

and that for any $\nu > 0$, for any $|u|$ sufficiently large depending on ν,

$$\begin{pmatrix} u \\ \nu|u| \end{pmatrix} \in H \text{ and } \lim_{|u| \to \infty} \frac{|u|}{U(u, \nu\,|u|)} = 0. \tag{1.18}$$

An explanation of how an entropy density U satisfying (1.17) and (1.18) can arise is provided in theorem 1.4 below.

THEOREM 1.1. *Assume a sequence $w^\epsilon = \binom{u^\epsilon}{v^\epsilon}$, $\epsilon \downarrow 0$, of approximate solutions of (1.7), such that w^ϵ and $U(w^\epsilon)$ are both bounded in L_1^{loc} uniformly with respect to ϵ pointwise in t, and such that the weak limit as $\epsilon \downarrow 0$ is of the form w^0 given in $(3.20)_1$. Assume U satisfies (1.13-1.18); then for all i, t, $M_i(t)$ is parallel to the vector $\binom{0}{1}$.*

REMARKS. In view of $(3.59)_1$, this implies that the first equation in (1.9), but not the second, will be satisfied with s the speed of propagation and $\binom{u}{v}$, $\binom{u_0}{v_0}$ the limiting values on the two sides of the singularity, as shown for the model problem in section 4 of chapter 1.

The set N may be empty, i.e., with $\overline{v}(u) = \infty$, as is the case for the model problem in $(3.2)_1$.

The proof uses a lemma, the proof of which is briefly deferred.

LEMMA 1.2. *Under the assumption (1.18), there exists $\underline{v}(u)$, for each $u \in \mathbb{R}$ such that*

$$\lim_{|u| \to \infty} \frac{|u|}{\underline{v}(u)} = 0, \tag{1.19}$$

and

$$\lim_{|u|\to\infty} \frac{|u|}{U(u,\underline{v}(u))} = 0. \tag{1.20}$$

PROOF OF THEOREM 1.1: It suffices to consider $w^\epsilon(\cdot,t)$ restricted to intervals J_ϵ, the length of which decreases to zero as $\epsilon \downarrow 0$, and such that

$$\int_{J_\epsilon} w^\epsilon(x,t)dx \overset{\epsilon\downarrow 0}{\to} M_i(t). \tag{1.21}$$

We decompose J_ϵ and the left side of (1.21)

$$J_\epsilon = \underline{J_\epsilon} \cup \overline{J_\epsilon},\ \underline{J_\epsilon}(\text{ resp. } \overline{J_\epsilon}) = \{x \in J_\epsilon \mid v^\epsilon(x,t) \le (\text{ resp } >)\underline{v}(u^\epsilon(x,t))\} \tag{1.22}$$

and designate

$$\underline{M_\epsilon}(\text{ resp. } \overline{M_\epsilon}) = \int_{\underline{J_\epsilon}(\text{ resp. } \overline{J_\epsilon})} w^\epsilon(x,t)dx. \tag{1.23}$$

Since the Lebesgue measures of $\underline{J_\epsilon}$ and $\overline{J_\epsilon}$ decrease to zero as $\epsilon \downarrow 0$, it is clear that only large values of $|w^\epsilon|$ can contribute to $\underline{M_\epsilon}$ and $\overline{M_\epsilon}$. Within $\overline{J_\epsilon}$, wherever $|u|$ becomes large one has

$$|u^\epsilon(x,t)| << \underline{v}(u^\epsilon(x,t)) < v^\epsilon(x,t)$$

and $v^\epsilon(\cdot,t) \in L_1^{loc}$ uniformly with respect to ϵ, so

$$\overline{M_\epsilon^{(1)}} \overset{\epsilon\downarrow 0}{\to} 0 \text{ and } \liminf_{\epsilon\downarrow 0} \overline{M_\epsilon^{(2)}} \ge 0, \tag{1.24}$$

superscripts denoting the two components of $\overline{M_\epsilon}$. Thus $\overline{M_\epsilon}$ is parallel to $\binom{0}{1}$.

Within $\underline{J_\epsilon}$, we have from (1.20), (1.15) and (1.22), again wherever $|u|$ becomes large

$$|u^\epsilon(x,t)| << U(u^\epsilon(x,t),\underline{v}(u^\epsilon(x,t))) \le U(u^\epsilon(x,t),v^\epsilon(x,t)) \tag{1.25}$$

so $U(w^\epsilon(\cdot,t)) \in L_1^{loc}$ uniformly in ϵ implies

$$\lim_{\epsilon\downarrow 0} \underline{M_\epsilon^{(1)}} = 0. \tag{1.26}$$

From (1.24) and (1.26),

$$M_i^{(1)}(t) = 0. \tag{1.27}$$

Finally, we use (1.17) to infer that

$$\left| \int_{\underline{J_\epsilon} \cap v^\epsilon(x,t)<0} v^\epsilon(x,t)dx \right| << \int_{\underline{J}_\epsilon(x,\epsilon)<0\cap v^\epsilon} U(u^\epsilon(x,t),v^\epsilon(x,t))dx \le c \tag{1.28}$$

so

$$\liminf_{\epsilon\downarrow 0} \underline{M_\epsilon^{(2)}} \ge 0 \tag{1.29}$$

and the proof of the theorem is reduced to that of lemma 1.2. □

PROOF OF LEMMA 1.2: By hypothesis (1.18), for any $\nu > 0$ (large), $\delta > 0$ (small), there exists $\kappa(\nu,\delta)$ satisfying

$$\frac{|u|}{U(u,\nu(u))} < \delta \text{ for all } |u| > \kappa(\nu,\delta). \tag{1.30}$$

Choose δ depending on ν such that

$$\delta(\nu) \to 0 \text{ as } \nu \to \infty \tag{1.31}$$

and ν depending on $|u|$ such that

$$\kappa(\nu(|u|), \delta(\nu(|u|))) = |u|. \tag{1.32}$$

From (1.30, (1.31) and (1.32),

$$\nu(|u|) \to \infty \text{ as } |u| \to \infty. \tag{1.33}$$

Finally we set

$$\underline{v}(u) = |u|\, \nu(|u|), \tag{1.34}$$

and (1.19) is immediate from (1.34). From (1.30), (1.33) and (1.35)

$$\frac{|u|}{U(u, \underline{v}(u))} = \delta(\nu(|u|)) \tag{1.35}$$

which approaches zero as $|u| \to \infty$ from (1.31) and (1.33). □

The assumptions of w^ϵ, $U(w^\epsilon) \in L_1^{loc}$ uniformly with respect to ϵ in theorem 1.1, for a system which is not everywhere hyperbolic, is based on the following theorem.

THEOREM 1.3. *Assume U smooth in w satisfying (1.14), (1.15), (1.17) and (1.18) in H, (1.16) in N, and for $w \in \partial H$, the boundary of H,*

$$U(w) = 0, \tag{1.36}$$

and

$$U_w(w) = 0. \tag{1.37}$$

Given $w_\pm \in \mathbb{R}^2$, $w(\cdot, 0) \in L_\infty : \mathbb{R} \to \mathbb{R}^2$, assume that w^ϵ, $\epsilon > 0$, satisfying

$$w_t^\epsilon + q(w_x^\epsilon) = \epsilon w_{xx}^\epsilon, \; x \in \mathbb{R}, \; t > 0 \tag{1.38}$$

$$w^\epsilon(\pm\infty, t) = w_\pm, \; t \geq 0 \tag{1.39}$$

$$w^\epsilon(x, 0) = w(x, 0), \; x \in \mathbb{R}, \; \epsilon > 0 \tag{1.40}$$

satisfy

$$\int\limits_{|x| > L(1+t)} |U(w^\varepsilon(x,t)) - U(w(x,0))| dx \leq c, \tag{1.41}$$

for some constants c, L independent of ε, t. Then w^ϵ, $U(w^\epsilon) \in L_1^{loc}(\mathbb{R})$ uniformly with respect to ϵ, pointwise in t.

REMARKS. A more general viscosity matrix can be used, replacing the right side of (1.38) by $\epsilon(Aw_x^\epsilon)_x$ with H such that $U_{ww}A > 0$ for all $w \in H$.

If $w_+ = w_-$, the assumption (1.41) is not needed. This seems a less interesting case, as we wish to allow a net input in w and/or $U(w)$.

There are various alternatives to the assumption (1.41) leading to the same conclusions.

PROOF. For $K > 0$, set

$$U_K(x) = \begin{cases} 0, & |x| \leq K \\ U(w(x,0)), & |x| > K. \end{cases} \tag{1.42}$$

Denote by

$$\mathcal{H} = \{x \mid w^\epsilon(x,t) \in H\} \tag{1.43}$$
$$\mathcal{N} = \{x \mid w^\epsilon(x,t) \in N\}. \tag{1.44}$$

While $\mathcal{H}, \mathcal{N}$ obviously depend on ϵ, t, this dependence will not be important and we drop the indicated subscripts for simplicity.

Multiplying (1.39) by $U_w(w^\epsilon)$ we have the usual entropy equation, which now holds only in $\mathcal{H}$

$$\begin{aligned} U(w^\epsilon)_t + F(w^\epsilon)_x &= \epsilon U_w(w^\epsilon) w^\epsilon_{xx} \\ &= \epsilon(U_w(w^\epsilon)w^\epsilon)_x - \epsilon U_{ww}(w^\epsilon) w^\epsilon_x w^\epsilon_x. \end{aligned} \tag{1.45}$$

where F is the entropy flux associated with U, satisfying

$$U_w(w) q_w(w) = F_w(w), \text{ for all } w \in H. \tag{1.46}$$

Using (1.37) and (1.46), we have $F_w \mid_{\partial H} = 0$, so F is constant on $\partial\mathcal{H}$ and without loss of generality

$$F \mid_{\partial H} = 0. \tag{1.47}$$

Next we use (1.36) and (1.45) to obtain, for any K,

$$\begin{aligned} \frac{d}{dt} \int_{\mathbb{R}} (U(w^\epsilon) - U_K)\, dx &= \int_{\mathcal{H}} U(w^\epsilon)_t dx \\ = \int_{\mathcal{H}} [-F(w^\epsilon)_x &- \epsilon U_{ww}(w^\epsilon) w^\epsilon_x w^\epsilon_x + \epsilon(U_w(w^\epsilon) w^\epsilon_x)_x]\, dx. \end{aligned} \tag{1.48}$$

At each t, the region $\mathcal{H}$ is a countable union of disjoint intervals, the endpoints of which are either at infinity, where w^ϵ_x vanishes and $w = w\pm$, or else at points at which $w \in \partial\mathcal{H}$, where F vanishes from (1.47) and U_w vanishes from (1.37).

We thus see that the right side of (1.48) does not exceed $F(w_-) - F(w_+)$; integrating with respect to t we find

$$\int_{\mathbb{R}} (U(w^\epsilon(x,t) - U_K(x)))\, dx \leq t(F(w_-) - F(w_+)). \tag{1.49}$$

We use (1.49) to show that $U(w^\epsilon) \in L_1^{loc}$ uniformly with respect to ϵ, pointwise in t. Suppose not: then there is a finite interval J and a $t_1 > 0$ such that

$$\int_J U(w^\epsilon(x,t_1)) dx \to \infty \text{ as } \epsilon \downarrow 0. \tag{1.50}$$

In (1.49) we choose $K \geq L(1+t_1)$, L the constant in (1.41), and such that $J \subset \{|x| < K\}$. Now using (1.49), (1.41) and (1.42), and the nonnegativity of U,

$$\begin{aligned}\int_J U(w^\epsilon(x,t_1))dx &= \int_J (U(w^\epsilon(x,t_1)) - U_K(x))\,dx \\ &= \int_{\mathbb{R}} (U(w^\epsilon(x,t_1)) - U_K(x))dx \\ &- \int_{|x|>K} (U(w^\epsilon(x,t_1)) - U_K(x))\,dx \\ &- \int_{|x|<K \quad x\notin J} (U(w^\epsilon(x,t_1) - U_K(x))dx \\ &\leq t_1(F(w_-) - F(w_+)) + c \qquad (1.51)\end{aligned}$$

contradicting (1.50).

Now with $w, q(w)$ as shown in (1.8), the second equation in (1.38) is

$$(v^\epsilon - v(\cdot,0))_t = -g(w^\epsilon)_x + \epsilon v^\epsilon_{xx}. \qquad (1.52)$$

Integrating over $\mathbb{R} \times (0,t)$ we find

$$\int_{\mathbb{R}} (v^\epsilon(x,t) - v(x,0))dx = -t(g(w_+) - g(w_-)). \qquad (1.53)$$

With U nonnegative and satisfying (1.17), there is a constant c depending on $v(\cdot,0)$ such that

$$|v^\epsilon(x,t) - v(x,0)| \leq v^\epsilon(x,t) - v(x,0) + c(U(w^\epsilon(x,t)) + 1). \qquad (1.54)$$

From (1.54), (1.41), (1.53) and (1.49) we have $v^\epsilon(\cdot,t) - v(\cdot,0) \in L_1$ uniformly with respect to ϵ, pointwise in t.

Stronger local estimates apply to u^ϵ, perhaps not surprisingly in view of the result of theorem 1.1.

Within $\mathcal{N}$, where $v^\epsilon(x,t) \geq \overline{v}(u^\epsilon(x,t))$, $v^\epsilon \in L_1^{loc}$ implies $\overline{v}(u^\epsilon) \in L_1^{loc}$, a stronger condition than $u^\epsilon \in L_1^{loc}$.

Appealing to lemma 1.2, in a region where $\underline{v}(u^\epsilon(x,t)) \leq v^\epsilon(x,t) < \overline{v}(u^\epsilon(x,t))$, $v^\epsilon \in L_1^{loc}$ implies $\underline{v}(u^\epsilon) \in L_1^{loc}$, again stronger than $u^\epsilon \in L_1^{loc}$ in view of (1.19).

Finally in a region where $v^\epsilon(x,t) < \underline{v}(u^\epsilon(x,t))$, from (1.15) we have

$$U(u^\epsilon(x,t), \underline{v}(u^\epsilon(x,t)) < U(u^\epsilon(x,t), v^\epsilon(x,t)) \qquad (1.55)$$

so $U(w^\epsilon) \in L_1^{loc}$ implies $U(u^\epsilon, \underline{v}(u^\epsilon)) \in L_1^{loc}$. This is also stronger than $u^\epsilon \in L_1^{loc}$ in view of (1.20). □

Systems of conservation laws admitting singular shock solutions have occasionally arisen in applications as a result of an exchange of conserved quantities [Ke]. This is at best a controversial procedure [S3]. For example, under an exchange of conserved quantities, the qualitative structure of the Hugoniot locus $\Gamma(w_0)$ is in general not maintained [S2]. The effect of such an interchange on the symmetry group for a given system was discussed in section 6 of chapter 1.

Such an exchange of conserved quantities provides a simple mechanism by which an entropy density U satisfying the assumptions of theorem 1.3 is obtained.

THEOREM 1.4. *Assume a system of the form*

$$u_t + f(u,\rho)_x = 0 \tag{1.56}$$

$$\rho_t + h(u,\rho)_x = 0 \tag{1.57}$$

for $(u,\rho) \in D = \{u \in \mathbb{R},\ \rho > 0\}$ *is equipped with a strictly convex (upwards) entropy density.*

$$v(u,\rho) = v_1(\rho) + v_2(u) \tag{1.58}$$

with

$$v_{1,\rho}(\mathbb{R}_+) = \mathbb{R}_- \tag{1.59}$$

and

$$v_{2,u}(\mathbb{R}) = \mathbb{R}. \tag{1.60}$$

Let $g = g(u,\rho)$ *denote the corresponding entropy flux, and denote by*

$$H = \{(u, v(u,\rho)),\ u \in \mathbb{R},\ \rho > 0\}. \tag{1.61}$$

Then the system

$$u_t + f(u,\rho(u,v))_x = 0 \tag{1.62}$$

$$v_t + g(u,\rho(u,v))_x = 0 \tag{1.63}$$

is equipped with a strictly convex (upwards) entropy density

$$U(u,v) = \left\{ \begin{array}{ll} \rho(u,v), & (u,v) \in H \\ 0, & (u,v) \in N \stackrel{def}{=} \mathbb{R}^2 - H \end{array} \right\} \tag{1.64}$$

with corresponding entropy flux

$$F(u,v) = \left\{ \begin{array}{ll} h(u,\rho(u,v)), & (u,v) \in H \\ 0 & (u,v) \in N \end{array} \right\} \tag{1.65}$$

satisfying (1.14-1.18), (1.36-1.37).

REMARKS. The assumed form of v (1.58) is convenient, but not essential to obtain such results; the examples in the next section are of this form. In any event, v and U are presumably not uniformly convex.

PROOF. From (1.59), for $u \in \mathbb{R},\ \rho > 0$

$$v_\rho(u,\rho) = v_{1,\rho}(\rho) < 0 \tag{1.66}$$

so the exchange of ρ with v as conserved quantities is justified [S2]. In particular H as designated in (1.61) satisfies

$$H = \{v < \overline{v}(u)\} \tag{1.67}$$

with $\overline{v}(u)$ determined from

$$\rho(u,v) \downarrow 0 \text{ as } v \uparrow \overline{v}(u). \tag{1.68}$$

Adding a constant to v if necessary, it is no loss of generality to assume $\overline{v}(u)$ nonnegative, thus recovering the partition of $\mathbb{R}^2$ shown in (1.13). For some or all $u, \overline{v}(u)$ may be infinite; indeed N may be empty.

From (1.68), $\rho(H) = \mathbb{R}_+$, so f, g in (1.62), (1.63) are defined for all $(u,v) \in H$. We assume f, g extended smoothly to N; then identifying $w = \binom{u}{v}$ the pair (1.62), (1.67) assumes the form (1.7), (1.8).

The conditions (1.59), (1.60) imply limits on the first partial derivatives of v

$$v_u(\pm\infty,\rho) = v_{2,u}(\pm\infty) = \pm\infty; \tag{1.69}$$

$$v_\rho(u,0+) = v_{1,\rho}(0+) = -\infty; \tag{1.70}$$

$$v_\rho(u,\infty) = v_{1,\rho}(\infty) = 0. \tag{1.71}$$

Again adding a constant to v if necessary, it is no loss of generality to assume

$$v_2(u) \geq 0, \tag{1.72}$$

and

$$v_1(0+) \geq 0; \tag{1.73}$$

indeed $v_1(0+)$ may be infinite.

As obtained in (1.64), U clearly satisfies (1.14), (1.16), (1.36); (1.15) follows immediately from (1.66). Now (1.37) follows from (1.36) and (1.70), in particular $U = \rho = 0$ on ∂H.

From (1.58), (1.64)

$$\begin{aligned} \liminf_{v\to-\infty} \frac{v}{U(u,v)} &= \liminf_{\rho\to\infty} \frac{v_1(\rho)+v_2(u)}{\rho} \\ &= \lim_{\rho\to\infty} \frac{v_1(\rho)}{\rho} \\ &= 0 \end{aligned} \tag{1.74}$$

by application of L'Hopital's rule, using (1.71), thus recovering (1.17).

From (1.68), (1.58) and (1.69)

$$\begin{aligned} \overline{v}_u(u) &= v_{2,u}(u) \\ &\to \pm\infty \text{ as } u \to \pm\infty \end{aligned} \tag{1.75}$$

so for any given ν, $\nu\,|u| \leq \overline{v}(u)$ for $|u|$ sufficiently large and $(u, \nu\,|u|) \in H$ for all sufficiently large $|u|$. Thus to obtain (1.18) we consider u/ρ on a curve

$$v(u,\rho) = \nu u \tag{1.76}$$

for ν fixed and u sufficiently large that $(u,\nu u) \in H$ and

$$v_{2,u}(u) > \nu \tag{1.77}$$

appealing to (1.69). (The case of large negative u is entirely similar).

Using (1.58) in (1.76) we have

$$-v_1(\rho) = v_2(u) - \nu u \tag{1.78}$$

so $\rho \to \infty$ as $u \to \infty$ satisfying (1.76) and

$$\frac{\partial\rho}{\partial u} = -\frac{v_{2,u}(u)-\nu}{v_{1,\rho}(\rho)} \to \infty \tag{1.79}$$

using (1.69) and (1.71). Again via L'Hopital's rule, (1.18) follows from (1.79). □

Given a system of the form (1.7), (1.8), equipped with an entropy density U satisfying (1.12-1.18), under additional conditions there exist values of $w_\pm \in \mathbb{R}^2$, $s \in \mathbb{R}$ such that there exist solutions of (1.38), (1.39) satisfying

$$w^\epsilon_{\underline{M}}(x,t) \overset{\epsilon\downarrow 0}{\to} \begin{cases} w_-, & x - st < 0 \\ w_+, & x - st > 0 \end{cases} + \begin{pmatrix} 0 \\ \underline{M} \end{pmatrix} \delta(x - st) \tag{1.80}$$

for any $\underline{M} > 0$. Identifying the independent variable in (1.4) as

$$\xi = \frac{x - st}{\epsilon} \tag{1.81}$$

we obtain

$$w^{\epsilon}_{\underline{M}}(x,t) = W^{\epsilon}_{\underline{M}}(\frac{x - st}{\epsilon}), \tag{1.82}$$

with W^{ϵ}_{M} a complete orbit of (1.4) (with $A = I$). The values of w_+, w_-, s are chosen so that

$$W^{\epsilon}_{\underline{M}}(\xi) \to w_{\pm} \text{ as } \xi \pm \infty \text{ and} \tag{1.83}$$

$$\epsilon \int_{-\infty}^{0} (W^{\epsilon}_{\underline{M}}(\xi) - w_-)d\xi + \epsilon \int_{0}^{\infty} (W^{\epsilon}_{\underline{M}}(\xi) - w_+)d\xi \to \begin{pmatrix} 0 \\ \underline{M} \end{pmatrix} \text{ as } \epsilon \downarrow 0. \tag{1.84}$$

From (1.83) and (1.4), for these solutions both Rankine-Hugoniot conditions (1.9) are satisfied, with $w_+ = (u, v)$ and $w_- = (u_0, v_0)$. The singular mass in (1.80) is thus independent of t, as expected from $(3.59)_1$. These solutions are thus a special class of singular shocks, often identified as "overcompressive shocks". Our objective here is to show that such weak limits of w^{ϵ}_{M} exist without reference to the specific definition of singular shock (1.3). The generalization to time-dependent singular mass, obtained by approximate solution of (1.38), will be made in section 3 below.

We continue with a needed lemma, an additional implication of the assumed features of the entropy density.

LEMMA 1.5. *Assume U smooth and strictly convex in a connected region $H \subseteq \mathbb{R}^2$ satisfying (1.36), (1.37). Assume a viscosity matrix $A = A(w)$ which is everywhere nonsingular and satisfies*

$$U_{ww}(w)A(w) > 0 \textit{ for all } w \in H. \tag{1.85}$$

Finally, assume any point w_0 such that the Hugoniot locus $\Gamma(w_0)$ satisfies

$$\Gamma(w_0) \subset H. \tag{1.86}$$

Then there are no nontrivial periodic orbits of (1.4) for any $s \in \mathbb{R}$, for this value of w_0.

PROOF. Consider a periodic orbit within N, the (closed) complement of H. Such an orbit would have to surround a critical point, necessarily a point in $\Gamma(w_0)$. As such critical points for (1.4) are in H from (1.86), and as H is connected, by hypothesis, such an orbit is impossible.

Therefore such an orbit must include a finite open interval J, such that from (1.36), (1.37) we have

$$w \mid_J \subset H, \; w \mid_{\partial J} \subset \partial H \tag{1.87}$$

$$U(w) \mid_{\partial J} = 0, \; U_w(w) \mid_{\partial J} = 0. \tag{1.88}$$

Differentiating (1.4) with respect to ξ, multiplying by $U_w(w(\xi))$ and integrating over J, we find

$$\begin{aligned} \int_J U_w(Aw')'d\xi &= \int_J (U_w q_w w' - sU_w w')d\xi \\ &= \int_J (F_w w' - sU_w w')d\xi, \end{aligned} \tag{1.89}$$

using (1.46) in the last step.

By partial integration, using (1.88), the left side of (1.89) is

$$-\int_J w' U_{ww} A w' d\xi < 0 \tag{1.90}$$

using (1.85). Using (1.46), the right side of (1.89) is

$$\int_J (F_w w' - sU_w w') d\xi = 0 \tag{1.91}$$

using (1.88) and (1.47) necessarily satisfied by the entropy flux F. The contradiction among (1.89), (1.90) and (1.91) completes the proof. □

We consider systems (1.7) admitting two closely related symmetries at infinity, as discussed in section 6, chapter 1. Specifically for $\epsilon > 0$, $(6.45 - 6.49)_1$ hold with

$$\tilde{Z}_\pm(\epsilon) = \begin{pmatrix} \pm\epsilon^{-k} & 0 \\ 0 & \epsilon^{-\ell} \end{pmatrix}, \quad \eta_{1\pm}(\epsilon) = \pm 1/\epsilon, \tag{1.92}$$

with the same

$$q^L(w) = \begin{pmatrix} f^L(u,v) \\ g^L(u,v) \end{pmatrix} \tag{1.93}$$

obtained for $\tilde{Z}_-(\epsilon)$ as for $\tilde{Z}_+(\epsilon)$, and with k, ℓ satisfying

$$0 < k \leq 1 < \ell. \tag{1.94}$$

Introducing polar coordinates α, ϕ, for phase space, with

$$u = \alpha^k \sin\phi \tag{1.95}$$

$$v = \alpha^\ell \cos\phi \tag{1.96}$$

from $(6.50)_1$, (1.92), (1.95), (1.96)

$$f^L(\pm\alpha^k \sin\phi,\ \alpha^\ell \cos\phi) = \alpha^{1+k} f^L(\sin\phi, \cos\phi), \tag{1.97}$$

$$g^L(\pm\alpha^k \sin\phi,\ \alpha^\ell \cos\phi) = \pm\alpha^{1+\ell} g^L(\sin\phi, \cos\phi), \tag{1.98}$$

in particular f^L is even and g^L odd with respect to ϕ.

With these assumptions, the condition (1.10) simplifies to

$$\sin\phi\, g^L(\sin\phi, \cos\phi) - \cos\phi\, f^L(\sin\phi, \cos\phi) \geq c > 0. \tag{1.99}$$

The functions f^L, g^L are nonetheless assumed such that there are solutions of

$$k\sin\phi\, g^L(\sin\phi, \cos\phi) - \ell\cos\phi\, f^L(\sin\phi, \cos\phi) = 0. \tag{1.100}$$

For ϕ satisfying (1.100), so does $-\phi$; $\phi = 0$ does not satisfy (1.100), using (1.99). Hereafter let ϕ_0 be the smallest positive solution of (1.100), assumed a simple zero satisfying

$$\phi_0 < \pi/2. \tag{1.101}$$

Systems satisfying these symmetry conditions admit singular shock solutions, as detailed in sections 3 and 4 below. Here we prove the existence of ordinary viscous structure for overcompressive shocks identified with such systems, but including nonzero singular mass.

THEOREM 1.6. *Assume a system (1.7) equipped with an entropy density* U *satisfying (1.14-1.18), (1.36), (1.37). Assume that within the region* H*, the system is strictly hyperbolic and genuinely nonlinear.*

Assume that the system satisfies two symmetries at infinity satisfying (1.92-1.94), and that (1.99), (1.100), and (1.101) hold.

Assume $w_0 \in H$ *given such that (1.86) holds, and that for any* $s \in \mathbb{R}$*, there are at most two nontrivial solutions of (1.9).*

Then there exist values of $s \in \mathbb{R}$*,* $w_\pm \in \Gamma(w_0)$ *such that there exists a solution of (1.4), (1.83), (1.84) for any* $\epsilon > 0, \underline{M} > 0$.

REMARKS. Using (1.82) and (1.81), this implies solutions of (1.38), (1.39) satisfying (1.80).

Throughout this proof A is the identity matrix.

The assumptions (1.100) and (1.101) can be replaced by additional assumptions on the entropy density. Also, the assumption of at most two nontrivial solutions of (1.9) can be relaxed [S5].

The form (1.7) determines u, v only up to an affine transformation. Much, but not all of the resulting ambiguity in u, v has been removed by the various assumptions made, in particular (1.15), (1.17), (1.18) and obviously (1.94). Nonetheless, in addition to scaling and/or additive constants in u, v all the results of this section survive transformations of the form $u \to u,\ v \to v + cu$ with arbitrary constant c. Indeed, singular shock systems often admit symmetries satisfying $(6.6)_1$ with $Z = \begin{pmatrix} 1 & 0 \\ c & 1 \end{pmatrix}$.

PROOF. For $|w|$ large, the trajectories determined by (1.4) (with $A =$ identity) are approximated by those of

$$\begin{aligned} u' &= f^L(u, v) \\ v' &= g^L(u, v). \end{aligned} \tag{1.102}$$

In the polar coordinates α, ϕ obtained in (1.95), (1.96), using (1.97), (1.98) the system (1.102) assume a simpler form

$$\alpha' = \alpha^2 \frac{\sin\phi\, f^L(\sin\phi, \cos\phi) + \cos\phi\, g^L(\sin\phi, \cos\phi)}{k \sin^2\phi + \ell \cos^2\phi} \tag{1.103}$$

$$\phi' = \alpha \frac{\ell \cos\phi\, f^L(\sin\phi, \cos\phi) - k \sin\phi\, g^L(\sin\phi, \cos\phi)}{k \sin^2\phi + \ell \cos^2\phi}. \tag{1.104}$$

Using (1.11), we find that $\theta = \tan^{-1}(v/u)$ satisfies

$$\begin{aligned} \theta' &= \alpha^{1+k+\ell} \frac{\sin\phi\, g^L(\sin\phi, \cos\phi) - \cos\phi\, f^L(\sin\phi, \cos\phi)}{\alpha^{2k} \sin^2\phi + \alpha^{2\ell} \cos^2\phi} \\ &> 0 \text{ for all } \alpha > 0 \end{aligned} \tag{1.105}$$

from (1.99). So θ is a Lyapunov function for the system (1.102), and the origin is the only critical point, of index $+1$.

From (1.103), α' is an odd function of ϕ; from (1.104), ϕ' is even in ϕ. Using (1.100), there are solutions of (1.103), (1.104) with $\phi = \pm\phi_0$ constant. For $\phi = \phi_0$

satisfying (1.101), $f^L(\sin\phi_0, \cos\phi_0)$ and $g^L(\sin\phi_0, \cos\phi_0)$ are of the same sign; both must be positive to satisfy (1.99) and (1.100). Thus

$$\alpha'(\alpha, \phi_0) > 0, \ \alpha'(\alpha, -\phi_0) < 0, \tag{1.106}$$

so the orbit corresponding to $\phi = \phi_0$ has the origin as its α-limit set and infinity as its ω-limit set. The opposite is true for the orbit corresponding to $\phi = -\phi_0$.

From (1.105) we have $f^L(0,1) < 0$, so $\phi'(\alpha, \phi) < 0, -\phi_0 < \phi < \phi_0$. It follows that all of the trajectories of (1.102) in the region determined by $\alpha > 0$, $|\phi| < \phi_0$ are complete homoclinic orbits, with the origin as both limit sets.

With w_0 fixed (and A = identity) in (1.4), we consider the trajectories $w(\xi, \mu, s)$, $-\infty < \xi < \infty$, determined by (1.4) with initial condition

$$w(0, \mu, s) = \begin{pmatrix} 0 \\ \mu \end{pmatrix} \tag{1.107}$$

with $\mu > 0$ large and $s \in \mathbb{R}$.

For $|w|$ large, the trajectories $w(\cdot, \mu, s)$ are approximated by the trajectories of (1.102) with initial data (1.107). Identifying u, v with the components of $w(\cdot, \mu, s)$ and determining α, ϕ from (1.95, 1.96), it follows that (1.103, 1.104) with lower order terms added hold for the system (1.4). In particular from (1.103, 1.104), using (1.99, 1.100) it follows that

$$\phi(w(\pm\mu^{-\frac{2}{3\ell}}, \mu, s)) \to \pm\phi_0 \text{ as } \mu \to \infty, \ s \text{ fixed.} \tag{1.108}$$

From (1.108), again comparing the trajectories of (1.4) and (1.102) for $|w|$ large, it follows that both limit sets of $w(\cdot, \mu, s)$ become independent of μ for μ sufficiently large depending on s.

For the system (1.4), the inequality (1.105) holds for α sufficiently large. In addition, lemma 1.5 precludes the existence of periodic orbits for the system (1.4). It follows that for each value of s, μ sufficiently large, at least one of the limit sets of $w(\cdot, \mu, s)$ is bounded, i.e. a single critical point.

Let Q_α (resp. Q_ω) denote the subset of $\mathbb{R}$ such that for $s \in Q_\alpha$ (resp. $s \in Q_\omega$) the α-limit set (resp. ω-limit set) of $w(\cdot, \mu, s)$ is bounded for all sufficiently large μ. Q_α and Q_ω are each not empty, containing all sufficiently small and large s, respectively, and their union is all of $\mathbb{R}$.

We claim that Q_α and Q_ω have a finite interval of intersection. To see this, we observe that as $\theta' > 0$ for sufficiently large $|w - w_0|$, from (1.105), it follows that $\Gamma(w_0)$ is bounded, i.e. the union of closed curves. Using (1.86) and strict hyperbolicity within H, it follows that there exist overcompressive shocks connecting w_0, i.e. there exists $s_0 \in \mathbb{R}$, $w_\pm \in \Gamma(w_0) \bigcup \{w_0\}$ such that $\lambda_+(w_+) < s_0 < \lambda_-(w_-)$, with the usual notation $\lambda_-(w) < \lambda_+(w)$ for all $w \in H$.

With $s = s_0$, the phase portrait of (1.4) has exactly three critical points, one of which is w_0. In particular, we identify w_- as the repeller and w_+ as the attractor. The third critical point is of index -1 and therefore a saddle.

Indeed, the phase portrait of (1.4) necessarily contains a single repeller for any $s < \lambda_-(w_-)$ and a single attractor for any $s > \lambda_+(w_+)$. Therefore by an elementary continuity argument, $(-\infty, \lambda_-(w_-)) \subset Q_\alpha$ and $(\lambda_+(w_+), \infty) \subset Q_\omega$. Finally $s = s_0 \in Q_\alpha \cap Q_\omega$, so with this value of s and μ sufficiently large, $w(\cdot, \mu, s)$ has w_- as its α-limit set and w_+ as its ω-limit set.

We can identify $W^\epsilon_{\underline{M}}$ satisfying (1.4), (1.83), (1.84) with $w(\cdot, \mu, s)$, for μ depending on $\underline{M}, \epsilon$, provided that (1.84) will be satisfied. Specifically, we need $|\underline{M}_1(\mu)|$

bounded and $\underline{M}_2(\mu) \to \infty$ as $\mu \to \infty$, where

$$\begin{pmatrix} \underline{M}_1(\mu) \\ \underline{M}_2(\mu) \end{pmatrix} = \int_{-\infty}^{0} (w(\xi, \mu, s) - w_-)d\xi + \int_{0}^{\infty} (w(\xi, \mu, s) - w_+)d\xi. \tag{1.109}$$

The potentially unbounded (with respect to μ) contributions to the right side of (1.109) occur for $|\xi|$ small, where $|w(\xi, \mu, s)|$ is large. For such ξ, the trajectories are approximated by those of (1.102), so it suffices to consider u, v given by (1.95), (1.96), α, ϕ satisfying (1.103), (1.104), with initial data

$$\alpha(0) = \mu^{1/\ell}, \ \phi(0) = 0. \tag{1.110}$$

For $|\phi|$ small and α large, from (1.103) and (1.104) we find

$$\phi' = \alpha(f^L(0, 1) + o(1)) \text{ as } \xi \to 0, \tag{1.111}$$

and

$$\left| \frac{d\alpha}{d\phi} \right| = O(\alpha), \tag{1.112}$$

so that

$$\alpha(\xi) \geq \mu^{1/\ell} e^{-c|\phi(\xi)|}. \tag{1.113}$$

Thus the leading term in $\underline{M}_1$ (μ) obtained in (1.109) is

$$\begin{aligned} \int \alpha^k(\xi) \sin\phi(\xi) d\xi \ &= \ \int \frac{\alpha^k \sin\phi}{|\phi'|} \, d\phi \\ &\leq \ c \int \alpha^{k-1} d\phi \end{aligned} \tag{1.114}$$

which is bounded using (1.94). As against this, the leading term in $\underline{M}_2(\mu)$ is

$$\begin{aligned} \int \alpha^\ell \cos\phi d\xi \ &= \ \int \frac{\alpha^\ell \cos\phi d\phi}{|\phi'|} \\ &\geq \ c \int \alpha^{\ell-1} d\phi \\ &\geq \ c\mu^{1-1/\ell} \end{aligned} \tag{1.115}$$

which indeed increases without bound with μ, again using (1.94). This completes the proof. □

2. Examples of systems admitting singular shocks

The best known examples of systems admitting singular shock solutions are closely related to the equations of isentropic gas dynamics in one space variable [Ke]

$$\rho_t + (\rho u)_x = 0 \tag{2.1}$$

$$(\rho u)_t + (\rho u^2 + \rho^\gamma)_x = 0. \tag{2.2}$$

The dependent variables in (2.1, 2.2) assume values in

$$D_0 = \{(\rho, \rho u) | \rho > 0, \quad u \in \mathbb{R}\} \tag{2.3}$$

and we assume in (2.2) that

$$1 \leq \gamma < 5/3. \tag{2.4}$$

The system (2.1, 2.2) is strictly hyperbolic, with characteristic speeds

$$\lambda_\pm(\rho, u) = \begin{cases} u \pm 1, & \gamma = 1 \\ u \pm (\gamma\rho^{\gamma-1})^{1/2}, & \gamma > 1. \end{cases} \tag{2.5}$$

Riemann invariants for these systems, satisfying

$$r_\pm \cdot \nabla \nu_\pm = 0 \tag{2.6}$$

where $r_\pm$ are the corresponding right eigenvectors, are given by

$$\nu_\pm(\rho, u) = u \mp \begin{cases} \log\ \rho, & \gamma = 1 \\ \frac{2}{\gamma-1}\left((\gamma\rho^{\gamma-1})^{1/2} - \gamma^{1/2}\right), & \gamma > 1. \end{cases} \tag{2.7}$$

With this choice of $\nu_\pm$, genuine nonlinearity follows from

$$\begin{aligned} \frac{\partial \lambda_\pm}{\partial \nu_\mp} &= \frac{1}{2}(1 + \frac{\gamma - 1}{2}) \\ &> 0. \end{aligned} \tag{2.8}$$

Smooth solutions of (2.1, 2.2) satisfy numerous additional conservation laws, two of which are

$$u_t + \left(\frac{u^2}{2} + \log\rho\right)_x = 0 \tag{2.9}$$

$$\left(\frac{u^2}{2} - \log\rho\right)_t + \left(\frac{u^3}{3} - u\right)_x = 0 \tag{2.10}$$

in the case $\gamma = 1$, and

$$u_t + \left(\frac{u^2}{2} + \frac{\gamma}{\gamma - 1}(\rho^{\gamma-1} - 1)\right)_x = 0 \tag{2.11}$$

$$\left(\frac{u^2}{2} - \frac{\gamma(\rho^{\gamma-1} - 1)}{(\gamma - 1)(2 - \gamma)}\right)_t + \left(\frac{u^3}{3} - \frac{\gamma}{2-\gamma}u\rho^{\gamma-1}\right)_x = 0 \tag{2.12}$$

in the case $\gamma > 1$.

In (2.7) ad (2.11, 2.12), arbitrary additive constants have been chosen so that the corresponding equations for the $\gamma = 1$ case are recovered in the limit $\gamma \downarrow 1$ with $\rho > 0$ fixed. Nevertheless the $\gamma = 1$ and $\gamma > 1$ cases are different in two important properties.

From (2.7), for any γ satisfying (2.4) it follows that the Riemann invariants $\nu_\pm$ are suitable global coordinates for phase space. For the case $\gamma = 1$, the region D_0 given by (2.3) corresponds to $\nu_\pm$ in all of $\mathbb{R}^2$, whereas for $\gamma > 1$, D_0 corresponds to the half plane

$$\nu_- - \nu_+ > -\frac{4\gamma^{1/2}}{\gamma - 1}, \tag{2.13}$$

obtained using (2.3) and (2.7).

The second distinction arises in the respective symmetry groups. In the notation of section 6, chapter 1, the system (2.1, 2.2) admits a reflection symmetry described by $(6.8)_1$, a Galilean symmetry described by $(6.9)_1$, and a scaling symmetry described by

$$Z(\alpha) = \begin{pmatrix} \alpha & 0 \\ 0 & \alpha^{\frac{\gamma+1}{2}} \end{pmatrix}, \eta_0 = \eta_2 = 0, \eta_3 = 1, \eta_1 = \alpha^{\frac{\gamma-1}{2}}; \alpha > 0, \gamma \geq 1. \tag{2.14}$$

For the case $\gamma = 1$, but not for $\gamma > 1$, the matrix $Z(\alpha)$ given by (2.14) commutes with $Z(\beta)$ given in $(6.9)_1$ for all $\alpha > 0, \beta \in \mathbb{R}$, and the scaling and Galilean symmetries determine a two parameter symmetry group. This greatly simplifies solution of the initial value problem for such systems, as is discussed beginning in section 5 below and in [KLS].

Using the primitive dependent variable v in (2.10) and in (2.12), determined from

$$v(u, \rho) = \begin{cases} \frac{1}{2}u^2 - \log \rho, & \gamma = 1 \\ \\ \frac{1}{2}u^2 - \frac{\gamma(\rho^{\gamma-1}-1)}{(\gamma-1)(2-\gamma)}, & \gamma > 1 \end{cases} \tag{2.15}$$

the system (2.9, 2.10) assumes the form

$$u_t + (u^2 - v)_x = 0 \tag{2.16}$$

$$v_t + \left(\frac{u^3}{3} - u\right)_x = 0 \tag{2.17}$$

whereas (2.11, 2.12) becomes

$$u_t + \left(\frac{3-\gamma}{2}u^2 - (2-\gamma)v\right)_x = 0 \tag{2.18}$$

$$v_t + \left(\frac{5-3\gamma}{6}u^3 + (\gamma-1)uv - \frac{\gamma u}{2-\gamma}\right)_x = 0. \tag{2.19}$$

We consider both systems (2.16, 2.17) and (2.18, 2.19) with $\binom{u}{v}$ assuming values in all of $\mathbb{R}^2$. For the system (2.16, 2.17), this corresponds to $\binom{\rho}{\rho u} \in D_0$, but represents an extension of phase space for the system (2.18, 2.19), for which $\binom{\rho}{\rho u} \in D_0$ given by (2.3) is equivalent to

$$\frac{1}{2}u^2 - v > -\frac{\gamma}{(\gamma-1)(2-\gamma)} \tag{2.20}$$

using (2.15), or to (2.13).

The system (2.16, 2.17) is the "model problem" for singular shocks, as discussed in sections 3 and 4 of chapter 1. With γ satisfying (2.4), the system (2.18, 2.19) also admits singular shock solutions. Here we show that each of these systems satisfies the assumptions of theorem 1.6, implying the existence of travelling wave solutions with nonzero singular mass, with viscous structure. The extension to true singular shock solutions, in which singular mass is accumulated with time, is made below in sections 3 and 4.

For these systems, the existence of entropy density/flux satisfying the requirements of theorem 1.6 is obtained by appeal to theorem 1.4. With fixed γ satisfying (2.4) a two-stage homotopy connects the system (2.1, 2.2) with (2.9, 2.10) or (2.11, 2.12). In the first stage, velocity u replaces momentum ρu as a conserved quantity, leading to a system consisting of (2.1) and (2.9) or (2.11). Smooth solutions with ρ

assuming positive values obviously survive this exchange; thus do the characteristic speeds and Riemann invariants. Genuine nonlinearity is maintained from (2.8).

In the second stage, (2.1) is exchanged with (2.10) or (2.12), i.e. v given by (2.15) replaces ρ as a primitive variable. With $u \in \mathbb{R}$ fixed, $\rho > 0$ and v satisfying (2.20) uniquely determine each other from (2.15), so again smooth solutions, characteristic speeds, Riemann invariants, and genuine nonlinearity survive the exchange [S2]. But the topological structure of the Hugoniot locus of a point in phase space does not [S2]; for the systems (2.16, 2.17) and (2.18, 2.19), $\Gamma(w_0)$ is bounded for any $w_0 \in \mathbb{R}^2$.

The function $v(u, \rho)$ given by (2.15) satisfies (1.58, 1.59, 1.60); the corresponding flux $g(u, \rho)$ is obtained from (2.10) or (2.12) by inspection. The region H, determined by (1.61), is all of $\mathbb{R}^2$ for $\gamma = 1$ and is the region determined by (2.13), or by (2.20), for $\gamma > 1$. The obtained entropy density/flux from (1.64, 1.65) is given by $(3.5)_1$ for the case $\gamma = 1$, and for $\gamma > 1$, u, v satisfying (2.20), by

$$U(u,v) = \Big[\frac{(\gamma-1)(2-\gamma)}{\gamma}\Big(\frac{1}{2}u^2 - v + \frac{\gamma}{(\gamma-1)(2-\gamma)}\Big)\Big]^{1/(\gamma-1)}, \tag{2.21}$$

$$F(u,v) = uU(u,v) \tag{2.22}$$

For both systems (2.16, 2.17) and (2.18, 2.19), the conditions (1.92-1.101) hold with $k = 1, \ell = 2$; the upper limit on γ shown in (2.4) is needed to verify (1.99).

For the system (2.16, 2.17), i.e. with H all of $\mathbb{R}^2$, the condition (1.86) is trivial. For the system (2.18, 2.19), for any $w_0 = \binom{u_0}{v_0} \in \mathbb{R}^2$, $\Gamma(w_0)$ is given explicitly by the set $\binom{u}{v}$ satisfying

$$\begin{aligned}\Big[\frac{(\gamma-1)}{4(2-\gamma)^{1/2}}(u-u_0)^2 - (2-\gamma)^{1/2}\Big(\frac{u^2-u_0^2}{2} - v + v_0\Big)\Big]^2 \\ + \frac{(\gamma+1)(5-3\gamma)}{48(2-\gamma)}(u-u_0)^4 \\ = (\gamma-1)\Big(\frac{u_0^2}{2} - v_0 + \frac{\gamma}{(\gamma-1)(2-\gamma)}\Big)(u-u_0)^2.\end{aligned} \tag{2.23}$$

For $\binom{u_0}{v_0} \notin H$, from (2.20) the right side of (2.23) is nonpositive, and no nontrivial solutions of (2.23) exist, using the limits (2.4) on γ. Thus (1.86) holds for the system (2.18, 2.19), for all $w_0 \in H$, and theorem 1.6 applies to this system.

The model problem (2.16, 2.17) admits an exceptional symmetry group, as described in Section 6 of chapter 1. This is not accidental. In the case $\gamma = 1$, both stages of the homotopy connecting the system (2.1, 2.2) with the system (2.9, 2.10) or (2.16, 2.17) satisfy the assumptions of theorem 6.3 of chapter 1 for the reflection, scaling and Galilean symmetries. Therefore the symmetry group for the system (2.17, 2.18) is inherited directly from that of (2.1, 2.2), with $\gamma = 1$.

This is not the case for $\gamma > 1$; only the reflection symmetry survives the exchanges of conserved quantities.

Singular shocks have also been identified for some systems for which theorem 1.6 does not hold. One such is obtained from the model problem (2.16, 2.17), replacing $q(w)$ with $q^L(w)$ obtained in (1.93) [SSS]

$$u_t + (u^2 - v)_x = 0, \tag{2.24}$$

$$v_t + \left(\frac{u^3}{3}\right)_x = 0, \quad \binom{u}{v} \in \mathbb{R}^2. \tag{2.25}$$

This system is not hyperbolic; u is a characteristic speed of multiplicity two, for which there exists only one corresponding right eigenvector. The single characteristic field is genuinely nonlinear, in contrast to the typical case for delta-shock systems. There is a single independent Riemann invariant,

$$\nu(u,v) = \frac{1}{2}u^2 - v. \tag{2.26}$$

Not being hyperbolic, the system (2.24, 2.25) cannot admit a convex entropy density. Indeed, let U, F denote an entropy density/flux for (2.24, 2.25), functions of u, ν with ν given by (2.26). Then U satisfies a heat equation

$$U_{uu} + U_\nu = 0 \tag{2.27}$$

and the corresponding F satisfies

$$F_u = u\,U_u, \quad F_\nu = U_u + u\,U_\nu. \tag{2.28}$$

For lack of a suitable entropy density, theorem 1.1 above cannot be applied to the system (2.24, 2.25). Nevertheless, the argument applied to the model problem in section 4 of chapter 1 applies equally well to the system (2.24, 2.25), and determines that any singular mass associated with a distribution solution must be parallel to $\binom{0}{1}$.

For any $w_0 = \binom{u_0}{v_0} \in \mathbb{R}^2$, the Hugoniot locus $\Gamma(w_0)$ is empty.

Until quite recently, the only systems known to admit singular shock solutions were pairs of conservation laws. The system

$$\begin{aligned} &u_t + (u^2 + \tilde{u}^2 - v)_x = 0 \\ &\tilde{u}_t + (u\tilde{u})_x = 0 \\ &v_t + \left(\frac{u^3}{3} + u\tilde{u}^2 - u\right)_x = 0 \end{aligned} \tag{2.29}$$

is strictly hyperbolic everywhere in $\mathbb{R}^3$, with characteristic speeds $u, u \pm (\tilde{u}^2+1)^{1/2}$. This system admits a strictly convex entropy density

$$U = e^{\frac{1}{2}(u^2+\tilde{u}^2)-v}, \quad F = uU \tag{2.30}$$

which satisfies an extended form of theorem 1.1 above, with u replaced by $(u, \tilde{u})$ assuming values in $\mathbb{R}^2$.

Solutions of the model problem (2.16, 2.17) are also solutions of (2.29) with $\tilde{u}$ vanishing identically, so it is clear that the system (2.29) admits some singular shock solutions. A larger class of singular shock solutions of (2.29) has been observed numerically [SS], with singular mass positive and only in the u-component as expected.

3. Viscous structure of singular shocks

The viscous structure of singular shocks is discussed here as an extension of the notion of the viscous structure of ordinary shocks. However, the analysis is complicated considerably by the lack of a traveling wave symmetry for singular shocks.

A brief review is perhaps helpful in explaining the situation. A jump discontinuity, w of the form

$$w(x,t) = \begin{cases} w_-, & x - st < 0 \\ w_+, & x - st > 0 \end{cases} \tag{3.1}$$

admits a traveling wave symmetry, as a function only of $x - st$, and is self-similar, as a function only of x/t. A weak solution of a system of conservation laws $(1.1)_1$ of the form (3.1) often is a weak limit of approximations obtained by regularization, replacing the flux vector $q(w)$ by

$$q^\varepsilon(w) = q(w) - \varepsilon A w_x \tag{3.2}$$

for some viscosity matrix A. The traveling wave symmetry is preserved by taking A constant; the self-similarity by taking A proportional to t [DD]. In either case, the available symmetry reduces the solution of the regularized system

$$w_t^\varepsilon + q(w^\varepsilon)_x - \varepsilon A w_{xx}^\varepsilon = 0, \quad x \in \mathbb{R},\ t > 0 \tag{3.3}$$

$$w^\varepsilon(\pm\infty, t) = w_\pm \tag{3.4}$$

to that of finding an orbit for a dynamical system connecting w_- to w_+ [F, DD].

The existence of such an orbit requires that w_-, w_+, s satisfy the usual Rankine-Hugoniot relations, thereby determining the Hugoniot locus of any given point in phase space and determining the shock speed s as a function of w_-, w_+. In this manner an entropy condition is also obtained, as the existence of such an orbit generally does not survive the interchange of w_- and w_+.

For weak shocks, with $|w_+ - w_-|$ sufficiently small, quite general conditions implying the existence and uniqueness of the solution of (3.3, 3.4) are known. As against that, for strong shocks, with which we are necessarily concerned here, results are much more limited [Liu4, Mo2].

For a singular shock, the weak limit of approximations corresponding to (3.1) is of the more general form

$$w^0(x,t) = \begin{cases} w_-, & x - st < 0 \\ w_+, & x - st > 0 \end{cases} + et\delta(x - st) \tag{3.5}$$

with $e = e(w_+, w_-, s)$ the (nonzero) Rankine-Hugoniot deficit,

$$e = s(w_+ - w_-) - q(w_+) + q(w_-). \tag{3.6}$$

The distribution w^0 given by (3.5) is of the form $(3.20)_1$; from theorem 3.3 of chapter 1, we have the corresponding weak limit of the flux functions given by

$$q^0(x,t) = \begin{cases} q(w_-), & x - st < 0 \\ q(w_+), & x - st > 0 \end{cases} + est\delta(x - st) \tag{3.7}$$

The limits w^0 and q^0 are both self-similar, depending only on x/t, but neither admits a traveling wave symmetry. Thus one could hope to obtain such limits of similarity solutions of (3.3, 3.4) with A proportional to t, but not with A constant. Such has been achieved numerically [SS], but not analytically, which is hardly surprising in view of the lack of known results for the corresponding problem for ordinary strong shocks.

An alternative is suggested by the observation that while neither w^0 nor q^0, given by (3.5) and (3.7), respectively, admits a traveling wave symmetry, the combination $sw^0 - q^0$ does. Thus $sw^0 - q^0$ may be sought as the weak limit of functions z^ε given by

$$z^\varepsilon(w^\varepsilon) = \varepsilon A w^\varepsilon_x - q(w^\varepsilon) + s w^\varepsilon \tag{3.8}$$

with A constant and the approximation w^ε admitting a traveling wave symmetry expressed in terms of z^ε,

$$z^\varepsilon(w^\varepsilon(x,t)) = z\left(\frac{x - st}{\varepsilon}\right). \tag{3.9}$$

In this context, we note that the function z obtained from (3.8, 3.9) is a constant in the application of (3.3), with A constant, to an ordinary shock of the form (3.1). The value of the constant z is, of course, determined by the asymptotic values (3.4); compatibility thereof requires that w_+, w_-, s satisfy the Rankine-Hugoniot relations, i.e. (3.6) with $e = 0$. With z so determined, (3.8) becomes an autonomous system of ordinary differential equations of dimension n, the familiar "viscous profile system".

For a singular shock, $e \neq 0$ in (3.6); from (3.8, 3.9, 3.3)

$$z(\infty) - z(-\infty) = e, \tag{3.10}$$

so z cannot be constant. Nevertheless one may choose a suitable, convenient form for z and then solve (3.8, 3.9), a first-order system of ordinary differential equations of dimension n, for w^ε. As w^0 given by (3.5) does not admit a traveling wave symmetry, i.e. is not a function only of $(x - st)$, a single trajectory of (3.8) for each w^ε will not suffice to obtain w^0 as the weak limit thereof. Thus w^ε is sought as a one-parameter family of trajectories of (3.8), (3.4). This characterizes the extension of the viscous structure for ordinary shocks to singular shocks: the non-constant function z in (3.9), and the determination of a one-parameter family of corresponding trajectories from (3.8, 3.4). This approach has been adopted in [SSS, S5].

The approximations w^ε so obtained do not satisfy (3.3) exactly; conditions are needed so that they satisfy (1.1-1.3) with w^0 given by (1.5). As regards the viscosity, in view of the discussion of section 1, only A a scalar multiple of the $n \times n$ identity matrix I_n will be considered here. The appeal to traveling wave symmetry (3.8, 3.9) then mandates the specific choice $A = I_n$. However, comparable results are obtained with "self-similar viscosity" $A = tI_n$, and this choice will also be discussed. The sharp division of existing literature on this point notwithstanding [KK2, KK3, SSS, S5], the choice between identity and self-similar viscosity seems to be remarkably unimportant.

For these two choices of viscosity, the following two theorems show how the conditions (1.1-1.3) are satisfied. We introduce two functions $y = y(\xi, \tau), z = z(\xi)$

for $-\infty < \xi < \infty$, $\tau \geq 0$, related by

$$z(\xi) = y_\xi(\xi, \tau) + sy(\xi, \tau) - q(y(\xi, \tau)). \tag{3.11}$$

THEOREM 3.1. *Assume the existence of y, z satisfying (3.11), and w^0 given by (3.5, 3.6) such that the following hold:*

$$z \in BV(\mathbb{R}), z(\pm\infty) = sw_\pm - q(w_\pm) \tag{3.12}$$

$$y(\cdot, 0) \in BV(\mathbb{R}) \tag{3.13}$$

$$y(\pm\infty, \tau) = w_\pm, \quad \tau \geq 0 \tag{3.14}$$

$$y_\tau(\cdot, \tau) \in L_1(\mathbb{R}), \quad \text{uniformly with respect to } \tau \tag{3.15}$$

$$\int\limits_{\mathbb{R}} y_\tau(\xi, \tau) d\xi = e + o(1) \quad \text{as} \quad \tau \to \infty \tag{3.16}$$

for all $\delta > 0$ there exists K_δ independent of τ such that

$$\int\limits_{|\xi| > K_\delta} |y_\tau(\xi, \tau)| d\xi \leq \delta. \tag{3.17}$$

Then

$$w^\varepsilon(x, t) = y\left(\frac{x - st}{\varepsilon}, \frac{t}{\varepsilon}\right) \tag{3.18}$$

satisfies (1.1-1.3), and z^ε obtained from (3.8) satisfies (3.9), with $A = I_n$ in (1.3) and in (3.18).

PROOF. With the identification of

$$\xi = \frac{x - st}{\varepsilon}, \quad \tau = \frac{t}{\varepsilon}, \tag{3.19}$$

and w^ε obtained from (3.18), the right side of (3.8) coincides with that of (3.11). Thus (3.9) is equivalent to z independent of τ, as shown in (3.11).

From (3.13, 3.14, 3.18) it follows that

$$w^\varepsilon(x, 0) \to w_+ \quad (\text{resp.} \quad w_-) \quad \text{for } x > 0 \quad (\text{resp.} \quad x < 0),$$

boundedly and pointwise in x as $\varepsilon \downarrow 0$. Thus again appealing to (3.13), both (1.1, 1.2) hold at $t = 0$.

Differentiating (3.18) and using (3.19) we find

$$\frac{\partial w^\varepsilon(x' + st, t)}{\partial t}\Big|_{x'} = \frac{1}{\varepsilon} y_\tau(\xi, \tau), \quad x' = x - st. \tag{3.20}$$

Thus for arbitrary fixed $t > 0$ and J an open bounded interval,

$$\begin{aligned} \int\limits_J |w^\varepsilon(x' + st, t) - w^\varepsilon(x', 0)| dx' &\leq \int\limits_J \Big| \int\limits_0^{t/\varepsilon} y_\tau\left(\frac{x'}{\varepsilon}, \tau\right) d\tau \Big| dx' \\ &= \varepsilon \int\limits_{\varepsilon\xi \in J} \int\limits_0^{t/\varepsilon} |y_\tau(\xi, \tau)| d\tau d\xi, \end{aligned} \tag{3.21}$$

using (3.19).

The condition (1.1) is now immediate from (3.21) and (3.15).

Next assume J bounded away from the point $x = 0$. Then for arbitrary small δ, we can choose $\varepsilon > 0$ sufficiently small that $x' \in J$ implies $|\xi| = |x'|/\varepsilon > K_\delta$ in (3.17). It follows that for such J, the right side of (3.21) approaches zero as $\varepsilon \downarrow 0$, and thus that $w^\varepsilon(\cdot + st, t) \to w_\pm$ as $\varepsilon \downarrow 0$ in $L_1(J)$.

For $x = 0$ within J, we again use (3.20) to obtain

$$\begin{aligned}\int_J (w^\varepsilon(x' + st, t) - w^\varepsilon(x', 0))dx' &= \int_J \int_0^{t/\varepsilon} y_\tau\Big(\frac{x'}{\varepsilon}, \tau\Big) d\tau dx' \\ &= \varepsilon \int_{\varepsilon\xi \in J} \int_0^{t/\varepsilon} y_\tau(\xi, \tau) d\tau d\xi \\ &= \varepsilon \int_0^{t/\varepsilon} \Big[\int_{\mathbb{R}} y_\tau(\xi, \tau) d\xi + o(1)\Big] d\tau \\ &= te + o(1) \qquad (3.22)\end{aligned}$$

as $\varepsilon \downarrow 0$, using (3.17) in the third step and (3.16) in the fourth. Now (1.2) follows from (3.22) and (3.5).

Differentiating (3.18) and using (3.11, 3.19) we find

$$w_t^\varepsilon + q(w^\varepsilon)_x - \varepsilon w_{xx}^\varepsilon = \frac{1}{\varepsilon}(y_\tau - z_\xi). \qquad (3.23)$$

Thus for arbitrary fixed $t > 0$ and θ continuous and of compact support, we have

$$\begin{aligned}\int_{\mathbb{R}} \theta(x - st)(w_t^\varepsilon(x,t) + q(w^\varepsilon(x,t))_x - \varepsilon w_{xx}^\varepsilon(x,t))dx &= \int_{\mathbb{R}} \theta(\varepsilon\xi)(y_\tau(\xi, \frac{t}{\varepsilon}) - z_\xi(\xi))d\xi \\ &= \theta(0)\Big[\int_{\mathbb{R}} y_\tau(\xi, \frac{t}{\varepsilon})d\xi - z(\infty) + z(-\infty)\Big] \\ &\quad + \int_{|\xi|<K} (\theta(\varepsilon\xi) - \theta(0))(y_\tau(\xi, \frac{t}{\varepsilon}) - z_\xi(\xi))d\xi \\ &\quad + \int_{|\xi|>K} (\theta(\varepsilon\xi) - \theta(0))(y_\tau(\xi, \frac{t}{\varepsilon}) - z_\xi(\xi))d\xi \qquad (3.24)\end{aligned}$$

for any value of K.

The first right-hand term is $o(1)$ as $\varepsilon \downarrow 0$ from (3.12) and (3.16); it is here that the specific form (3.6) for e is needed. By choosing K sufficiently large, the third right-hand term in (3.24) is made arbitrarily small, using (3.12) and (3.17). Then choosing ε sufficiently small depending on K, the second right-hand term in (3.24) becomes arbitrarily small using the continuity of θ, (3.12) and (3.15). Thus (1.3) is established. □

A similar result holds using self-similar viscosity.

THEOREM 3.2. *In addition to the hypotheses of theorem 3.1, assume y satisfying*

$$\int_{\mathbb{R}} |\xi y_\xi(\xi,0)|d\xi < \infty, \tag{3.25}$$

and

$$\int_{\mathbb{R}} |\xi y_\tau(\xi,\tau)|d\xi < \infty \text{ uniformly with respect to } \tau. \tag{3.26}$$

Then

$$\tilde{w}^\varepsilon(x,t) = y\Big(\frac{x-st}{\varepsilon t}, \frac{1}{\varepsilon}\Big) \tag{3.27}$$

satisfies (1.1), (1.2), and

$$\tilde{w}^\varepsilon_t(\cdot,t) + q(\tilde{w}^\varepsilon(\cdot,t))_x - \varepsilon t\tilde{w}^\varepsilon_{xx}(\cdot,t) \rightharpoonup 0 \text{ as } \varepsilon \downarrow 0 \tag{3.28}$$

weakly in the dual of the space of Lipshitz continuous functions of compact support.

REMARKS. The same function y is used to construct the approximations w^ε (3.18) and $\tilde{w}^\varepsilon$ (3.27). In this sense the different form of viscosity corresponds to an elementary change of variable.

The function $z^\varepsilon(w^\varepsilon)$ obtained from (3.8) and (3.27), with $A = t\,I_n$, is a function only of $(x/t - s)/\varepsilon$, in contrast to (3.9).

PROOF. Comparing (3.18) and (3.27) we observe that

$$\tilde{w}^{\varepsilon/t}(x,t) = w^\varepsilon(x,t), \;\; t > 0, \;\; x \in \mathbb{R}, \tag{3.29}$$

so as the w^ε satisfy (1.1, 1.2), so do the $\tilde{w}^\varepsilon$. Thus it suffices to prove (3.28). In this case we identity

$$\xi = \frac{x-st}{\varepsilon t}, \tag{3.30}$$

then differentiating (3.27) and using (3.11), (3.30) we find

$$\begin{aligned} &\tilde{w}^\varepsilon_t(x,t) + q\Big(\tilde{w}^\varepsilon(x,t)\Big)_x - \varepsilon t\tilde{w}^\varepsilon_{xx}(x,t) \\ &\quad = -\frac{1}{t}\Big(\xi y_\xi\Big(\xi, \frac{1}{\xi}\Big) + \frac{1}{\varepsilon}z_\xi(\xi)\Big). \end{aligned} \tag{3.31}$$

Thus for fixed $t > 0$ and θ Lipshitz continuous and of compact support,

$$\begin{aligned} &\int_{\mathbb{R}} \theta(x-st)(\tilde{w}^\varepsilon_t(x,t) + q(\tilde{w}^\varepsilon(x,t))_x - \varepsilon t\tilde{w}^\varepsilon_{xx}(x,t))dx \\ &\quad = -\varepsilon\int_{\mathbb{R}} \xi\theta(\varepsilon t\xi)y_\xi\Big(\xi, \frac{1}{\xi}\Big)d\xi - \int_{\mathbb{R}} \theta(\varepsilon t\xi)z_\xi(\xi)d\xi. \end{aligned} \tag{3.32}$$

The second right-hand term in (3.32) is evaluated as in the proof of theorem 3.1,

$$
\begin{aligned}
-\int_{\mathbb{R}} \theta(\varepsilon t\xi) z_\xi(\xi)d\xi = -\theta(0)\int_{\mathbb{R}} z_\xi(\xi)d\xi \\
- \int_{|\xi|>K} (\theta(\varepsilon t\xi)-\theta(0)) z_\xi(\xi)d\xi \\
- \int_{|\xi|<K} (\theta(\varepsilon t\xi)-\theta(0)) z_\xi(\xi)d\xi \\
= -\theta(0)e + o(1) \quad \text{as} \quad \varepsilon \downarrow 0,
\end{aligned}
\tag{3.33}
$$

using (3.12) and (3.16) in the first term, taking K large and using (3.12) in the second, and the continuity of θ and (3.12) in the third term.

The first right-hand term of (3.32) is equal to

$$
\begin{aligned}
&-\varepsilon \int_0^{1/\varepsilon}\int_{\mathbb{R}} \xi\theta(\varepsilon t\xi) y_{\xi\tau}(\xi,\tau)d\xi d\tau - \varepsilon\int_{\mathbb{R}} \xi\theta(\varepsilon t\xi)y_\xi(\xi,0)d\xi \\
&= \varepsilon \int_0^{1/\varepsilon}\int_{\mathbb{R}} \theta(\varepsilon t\xi) y_\tau(\xi,\tau)d\xi d\tau + \varepsilon^2 t \int_0^{1/\varepsilon}\int_{\mathbb{R}} \xi\theta_x(\varepsilon t\xi)y_\tau(\xi,\tau)d\xi d\tau \\
&\quad - \varepsilon\int_{\mathbb{R}} \xi\theta(\varepsilon t\xi)y_\xi(\xi,0)d\xi
\end{aligned}
\tag{3.34}
$$

after a partial integration. The third right-hand term in (3.34) is $O(\varepsilon)$ using (3.25), and the second term is also of order $O(\varepsilon)$ using (3.26), as the weak derivative θ_x is bounded in L_∞. The first term is equal to

$$
\begin{aligned}
&\varepsilon \int_0^{1/\varepsilon}\int_{\mathbb{R}} \theta(\varepsilon t\xi) y_\tau(\xi,\tau)d\xi d\tau \\
&= \varepsilon\theta(0) \int_0^{1/\varepsilon}\int_{\mathbb{R}} y_\tau(\xi,\tau)d\xi d\tau + \varepsilon \int_0^{1/\varepsilon}\int_{|\xi|>K} (\theta(\varepsilon t\xi)-\theta(0)) y_\tau(\xi,\tau)d\xi d\tau \\
&\quad K + \varepsilon \int_0^{1/\varepsilon}\int_{|\xi|<K} (\theta(\varepsilon t\xi)-\theta(0)) y_\tau(\xi,\tau)d\xi d\tau \\
&= \theta(0)e + o(1) \quad \text{as} \quad \varepsilon \downarrow 0,
\end{aligned}
\tag{3.35}
$$

using (3.16), (3.17), (3.15) and the continuity of θ, by the same argument used to obtain (3.33). Now from (3.33) and (3.35), the right side of (3.32) is $o(1)$ as $\varepsilon \downarrow 0$, establishing (3.28). □

The function y satisfying (3.11-3.17) and possibly (3.25, 3.26) is established by phase plane analysis for pairs of conservation laws satisfying the structural conditions described in section 1. Indeed, there are additional systems with quite different structure for which such results are obtained. But for each system of conservation laws admitting singular shocks, there are important conditions on the states $w_\pm$ which can be so connected and the corresponding speed s. Thus we now address the question of the "singular shock Hugoniot locus", denoted by $\Gamma_S(w_-)$, the set of states $w_+ \in D$ which can be connected on the right to a given state $w_- \in D$ on the left by a singular shock, and the corresponding speed of propagation. This requires a precise statement of what it means to connect two states by a singular shock.

DEFINITION. A state w_- is connected on the left to a state w_+ on the right by a singular shock of speed s if there exists a sequence w^ε satisfying (1.1-1.3), (3.8-3.9) with weak limit w° of the form

$$w^\circ(x,t) = \begin{cases} w_-, x - st < 0 \\ w_+, x - st > 0 \end{cases} + (M_0 + te)\delta(x - st), \quad 0 < t < \bar{t} \tag{3.36}$$

with e given by (3.6), e and/or some $M_0 \in \mathbb{R}^n$ nonzero, and some $\bar{t} > 0$.

This definition is somewhat arbitrary in the form of convergence specified in (1.3) and in the inclusion of the symmetry condition (3.8, 3.9). Indeed, these have been chosen so that we can prove the existence of precisely enough singular shocks to solve the Cauchy problem for a class of systems including the model problem (2.16, 2.17).

The following theorem, the principal result of this section, gives sufficient conditions on the flux $q(w)$, the points $w_\pm$ and the speed s for the existence of a singular shock connecting w_- to w_+, for a system of the form (1.7). The result is perhaps suspicious in its generality. The cognitions (1.13-1.18) on the entropy density are not needed. Strict hyperbolicity and genuine nonlinearity are needed only at each of the points $w_\pm$, and this only to permit a weakening of the overcompressibility condition. As against this, the required conditions on w_+, w_-, s may be more restrictive than is apparent. It is not immediately clear that such points exist; this will be clarified in section 5.

We need a technical assumption on the lower order parts of the flux functions. Denote these by

$$f^\circ(u,v) = f(u,v) - f^L(u,v) \tag{3.37}$$

$$g^\circ(u,v) = g(u,v) - g^L(u,v) \tag{3.38}$$

where the leading terms f^L, g^L satisfy (1.97, 1.98). From $(6.48)_1$, with u, v as in (1.95, 1.96), f°, g° satisfy

$$\lim_{\alpha\to\infty} \frac{f^\circ(\alpha^k \sin\phi, \alpha^\ell \cos\phi)}{\alpha^{1+k}} = \lim_{\alpha\to\infty} \frac{g^\circ(\alpha^k \sin\phi, \alpha^\ell \cos\phi)}{\alpha^{1+\ell}} = 0. \tag{3.39}$$

We require conditions on the first derivatives, of the form

$$\begin{aligned} &\lim_{\alpha\to\infty} \frac{(f_u^\circ(\alpha^k \sin\phi, \alpha^\ell \cos\phi)}{\alpha}, \lim_{\alpha\to\infty} \frac{f_v^\circ(\alpha^k \sin\phi, \alpha^\ell \cos\phi)}{\alpha^{1+k-\ell}}, \\ &\lim_{\alpha\to\infty} \frac{g_u^0(\alpha^k \sin\phi, \alpha^\ell \cos\phi)}{a^{1+\ell-k}}, \lim_{\alpha\to\infty} \frac{g_v^\circ(\alpha^k \sin\phi, \alpha^\ell \cos\phi)}{\alpha} = 0 \end{aligned} \tag{3.40}$$

which are satisfied, for example, by polynomials f°, g°.

THEOREM 3.3. *Assume a pair of conservation laws (1.7, 1.8) admitting two symmetries at infinity satisfying (1.92-1.94), (1.99-1.101), (3.39) and (3.40). Assume the system strictly hyperbolic and genuinely nonlinear at each of two points* $w_\pm$*, which together with a given* $s \in \mathbb{R}$ *satisfy a generalized Rankine-Hugoniot condition*

$$s(u_+ - u_-) - f(w_+) + f(w_-) = 0 \tag{3.41}$$

$$s(v_+ - v_-) - g(w_+) + g(w_-) \geq 0 \tag{3.42}$$

and a weak overcompressibility condition

$$\lambda_+(w_+) \leq s \leq \lambda_-(w_-) \tag{3.43}$$

Then there exist y, z *satisfying (3.11-3.17).*

REMARKS. In (3.43) $\lambda_\pm$ are the characteristic speeds, with $\lambda_-(w_\pm) < \lambda_+(w_\pm)$. The assumptions of hyperbolicity and genuine nonlinearity can be relaxed if (3.43) is replaced by

$$Re(\lambda_+(w_+)) < s < Re(\lambda_-(w_-)) \tag{3.44}$$

where $Re(\lambda_-(w_\pm)) \leq Re(\lambda_+(w_\pm))$.

The case where (3.42) holds with equality corresponds to an overcompressive shock with finite singular mass, $e = 0$ and $M_0 \neq 0$ in (3.36). This special case is treated separately. For (3.42) a strict inequality, by appeal to theorem 3.1, y, z satisfying (3.11-3.17) implies w^ε satisfying (1.1-1.3), and thus the existence of a singular shock of speed s connecting w_- to w_+ according to the definition given above. For this case in (3.36) we have

$$M_0 = \begin{pmatrix} 0 \\ m_0 \end{pmatrix}, \quad m_0 \geq 0, \quad \bar{t} = \infty \tag{3.45}$$

and

$$e = \begin{pmatrix} 0 \\ e_2 \end{pmatrix}, \quad e_2 > 0, \tag{3.46}$$

e_2 obtained as the left side of (3.42).

PROOF. It suffices to consider the system (3.11) with $\xi > 0$, as the argument for $\xi < 0$ is entirely similar. Initial conditions at $\xi = 0$ are taken of the form

$$y_1(0, \tau) = 0, \tag{3.47}$$

$$y_{2,\tau}(0, \tau) > 0, \quad y_2(0, \tau) \to \infty \text{ as } \tau \to \infty, \tag{3.48}$$

the precise condition on $y_2(0, \cdot)$ to be determined below.

The proofs of four lemmas are deferred to the next section in the interest of continuity of the argument. The first lemma shows that the large values of $|y|$ implied by (3.48) occur only in a neighborhood of $\xi = 0$.

LEMMA 3.4. *The limit*

$$p_0 \stackrel{def}{=} \lim_{\tau \to \infty} y(\xi_0, \tau) \tag{3.49}$$

exists for any sufficiently small $\xi_0 > 0$*,* $y(\cdot, \tau)$ *the solutions of (3.11), (3.47), 3.48) with* $z(\xi) = 0$*.*

Without loss of generality we take $\xi_0 \le 1$, and $z(\xi)$ of the form

$$z(\xi) = \begin{cases} 0, 0 < \xi \le \xi_0 \\ p_\xi(\xi) + sp(\xi) - q(p(\xi)),\ \xi_0 < \xi < 2 \\ sw_+ - q(w_+),\ \xi \ge 2 \end{cases} \tag{3.50}$$

where

$$p(\xi) = \left(\frac{2-\xi}{2-\xi_0}\right) p_0 + \left(\frac{\xi-\xi_0}{2-\xi_0}\right) p_1, \tag{3.51}$$

p_1 to be determined below.

In the interval $\xi_0 < \xi < 2$, the system (3.11), (3.50) is

$$(y(\xi,\tau) - p(\xi))_\xi = q(y(\xi,\tau)) - q(p(\xi)) - s(y(\xi,\tau) - p(\xi)), \tag{3.52}$$

so from (3.49) it follows that

$$\lim_{\tau\to\infty} y(2,\tau) = p_1. \tag{3.53}$$

For $\xi > 2$, the system (3.11), (3.50) is a standard "viscous profile" system,

$$y(\xi,\tau)_\xi = q(y(\xi,\tau)) - q(w_+) - s(y(\xi,\tau) - w_+). \tag{3.54}$$

Thus to satisfy (3.14) it suffices to choose p_1 an interior point of the stable manifold of w_+ for the system (3.54). For $s > \lambda_+(w_+)$ (or $s > Re(\lambda_+(w_+))$ if (3.44) holds) we can choose simply $p_1 = w_+$, as w_+ is an attractive node (or spiral).

For $s = \lambda_+(w_+)$ denote by $r_\pm$ the right eigenvectors corresponding to $\lambda_\pm(w_+)$, and $\kappa = \lambda_+(w_+) - \lambda_-(w_+)$. In a neighborhood of w_+ we employ an expansion

$$y(\xi,\tau) - w_+ = \zeta(\xi,\tau)r_+ + \omega(\xi,\tau)r_-. \tag{3.55}$$

LEMMA 3.5. *Assume the pair (1.7, 1.8) strictly hyperbolic and genuinely nonlinear at w_+. Then with $s = \lambda_+(w_+)$ and ζ, ω obtained from (3.55), in a neighborhood of w_+ the system (3.54) assumes the form*

$$\zeta_\xi(\cdot,\tau) = -\zeta(\cdot,\tau)^2 + h_1(\zeta(\cdot,\tau),\omega(\cdot,\tau)) \tag{3.56}$$

$$\omega_\xi(\cdot,\tau) = -\kappa\omega(\cdot,\tau) + h_2(\zeta(\cdot,\tau),\omega(\cdot,\tau)), \tag{3.57}$$

and there is a constant c_5 such that

$$|h_1(\zeta,\omega)| \le c_5(|\zeta|^3 + |\zeta\omega| + \omega^2) \tag{3.58}$$

$$|h_2(\zeta,\omega)| \le c_5(\zeta^2 + \omega^2) \tag{3.59}$$

$$|h_{1,\zeta}(\zeta,\omega)| \le c_5(\zeta^2 + |\omega|) \tag{3.60}$$

$$|h_{1,\omega}(\zeta,\omega)| \le c_5(|\zeta| + \omega) \tag{3.61}$$

$$|h_{2,\zeta}(\zeta,\omega)|\ ,\ |h_{2,\omega}(\zeta,\omega)| \le c_5(|\zeta| + |\omega|). \tag{3.62}$$

It will be shown in the proof of lemma 3.7 below that for all $c_6 > 0$ sufficiently small the point

$$p_1 = w_+ + c_6 r_+ \tag{3.63}$$

is an interior point of the stable manifold of w_+ for (3.54).

Thus for the case $s = \lambda_+$, the point p_1 is obtained from (3.63) with c_6 chosen sufficiently small below, and (3.14) holds in this case as well.

The form of $z(\xi)$ given in (3.50) obviously satisfies (3.12), so it suffices to prove (3.15-3.17). In the case where (3.42) holds with equality, we replace (3.48) by

$y_2(0,\tau)$ sufficiently large and independent of τ. Then $y(\cdot,\tau)$ is independent of τ and (3.15-3.17) are trivial. For (3.42) a strict inequality, we proceed by partitioning the various integrals in (3.15-3.17).

LEMMA 3.6. *The function $y_2(0,\tau)$ can be chosen satisfying (3.48) so that*

$$\int_0^{\xi_0} y_\tau(\xi,\tau)d\xi + \text{(corresponding term from } \xi<0) = \begin{pmatrix} o(1) \\ e_2 \end{pmatrix} \text{ as } \tau\to\infty, \tag{3.64}$$

$$\int_0^{\xi_0} |y_\tau(\xi,\tau)|d\xi < \infty \text{ uniformly in } \tau, \tag{3.65}$$

and

$$\lim_{\tau\to\infty} y_\tau(\xi_0,\tau) = 0. \tag{3.66}$$

From (3.49), (3.52) and (3.56),

$$\int_{\xi_0}^{2} |y_\tau(\xi,\tau)|d\xi \to 0 \tag{3.67}$$

and

$$y_\tau(2,\tau)\to 0 \tag{3.68}$$

as $\tau\to\infty$, so it suffices to consider the interval $\xi>2$, with $y(\cdot,\tau)$ satisfying (3.54) and (3.14).

To obtain (3.15), (3.16) and (3.17), it suffices to show that

$$\int_2^{\infty} |y_\tau(\xi,\tau)|d\xi \to 0 \text{ as } \tau\to\infty. \tag{3.69}$$

For the case $s>\lambda_+(w_+)$ (or $s>Re(\lambda_+(w_+))$, (3.69) follows easily from (3.68) and (3.53), as from (3.54) $|y(\cdot,\tau)-w_+|$ and $|y_\tau(\cdot,\tau)|$ decay exponentially for large ξ.

In the characteristic case, where $s=\lambda_+(w_+)$, we appeal to the following.

LEMMA 3.7. *For $s=\lambda_+(w_+)$ and $c_6>0$ sufficiently small, $y_\tau(\cdot,\tau)$ obtained from (3.54), (3.53), (3.63) satisfies (3.69) and (3.14).*

Thus the proof of theorem 3.3 is reduced to those of the four subsequent lemmas. □

4. Proofs of lemmas 3.4–3.7

PROOF OF LEMMA 3.4: We write the system (3.11) applied to the pair (1.7, 1.8) with $z=0$ in polar coordinates

$$\begin{aligned} y_1(\xi,\tau) &= -r(\xi,\tau)^k \sin\phi(\xi,\tau) \\ y_2(\xi,\tau) &= r(\xi,\tau)^\ell \cos\phi(\xi,\tau) \end{aligned} \tag{4.1}$$

with k,ℓ as appearing in (1.92-1.98). The leading and lower order terms in $q(w)$ are shown in (3.37, 3.38) with the leading terms f^L, g^L satisfying (1.97-1.98) and the lower order terms f^0, g^0 satisfying (3.39-3.40).

Using (4.1) in (3.11) in this form and using (1.97-1.98) explicitly we obtain an autonomous pair for r, ϕ,

$$r_\xi = -r^2 T_1(\phi) + R_1(r,\phi) \tag{4.2}$$

$$\phi_\xi = rT_2(\phi) + R_2(r,\phi) \tag{4.3}$$

with

$$T_1(\phi) = [\cos\phi g^L(\sin\phi,\cos\phi) + \sin\phi f^L(\sin\phi,\cos\phi)]/ (k\sin^2\phi + \ell\cos^2\phi) \tag{4.4}$$

$$T_2(\phi) = [k\sin\phi g^L(\sin\phi,\cos\phi) - \ell\cos\phi f^L(\sin\phi,\cos\phi)]/ (k\sin^2\phi + \ell\cos^2\phi) \tag{4.5}$$

$$R_1(r,\phi) = [-rs + r^{1-\ell}\cos\phi g^0(-r^k\sin\phi, r^\ell\cos\phi) - r^{1-k}\sin\phi f^0(-r^k\sin\phi, r^\ell\cos\phi)]/ (k\sin^2\phi + \ell\cos^2\phi) \tag{4.6}$$

$$R_2(r,\phi) = [-s(\ell-k)\sin\phi\cos\phi - kr^{-\ell}\sin\phi\, g^0(-r^k\sin\phi, r^\ell\cos\phi) - \ell r^{-k}\cos\phi\, f^0(-r^k\sin\phi, r^\ell\cos\phi)]/(k\sin^2\phi + \ell\cos^2\phi). \tag{4.7}$$

Initial conditions at $\xi = 0$ are obtained from (3.47-3.48), using (4.1)

$$r_\tau(0,\tau) > 0, \quad r(0,\tau) \to \infty \text{ as } \tau \to \infty \tag{4.8}$$

$$\phi(0,\tau) = 0. \tag{4.9}$$

The proofs of lemmas 3.4 and 3.6 are obtained by analysis of the solutions of (4.2-4.9), parameterized in τ. Using (3.39) and (3.40), the "remainder" terms R_1, R_2 are small for large r in the sense that

$$\lim_{r\to\infty}\Big[\frac{|R_1(r,\phi)|}{r^2} + \frac{|R_{1,r}(r,\phi)|}{r} + \frac{|R_{1,\phi}(r,\phi)|}{r^2}\frac{|R_2(r,\phi)|}{r} + |R_{2,r}(r,\phi)| + \frac{|R_{2,\phi}(r,\phi)|}{r}\Big] = 0. \tag{4.10}$$

Within a region devoid of critical points, a two-dimensional autonomous system such as (4.2, 4.3) can be reduced at least locally to a scalar ordinary differential equation. In the present case, the identification of appropriate independent variables depends on the behavior of the functions $T_1(\phi), T_2(\phi)$. From (1.100) and (1.101) we have $\phi_0 \in (0, \pi/2)$ such that

$$T_2(\phi_0) = 0. \tag{4.11}$$

Setting $\phi = 0$ in (1.99) we have $f^L(0,1) < 0$ and thus

$$T_2(\phi) > 0, \quad 0 \le \phi < \phi_0. \tag{4.12}$$

Solving (1.99) and (1.100) at $\phi = -\phi_0$ we obtain

$$(\ell-k)\cos\phi_0 f^L(\sin\phi_0,\cos\phi_0) \ge c \tag{4.13}$$

$$(\ell-k)\sin\phi_0 g^L(\sin\phi_0,\cos\phi_0) \ge c, \tag{4.14}$$

$c > 0$ the constant in (1.99).

We use (4.13) and (4.14) to estimate

$$
\begin{aligned}
&T_1(\phi_0)\\
&=\frac{\cos^2\phi_0\sin\phi_0 g^L(\sin\phi_0,\cos\phi_0)+\sin^2\phi_0\cos\phi_0 f^L(\sin\phi_0,\cos\phi_0)}{\sin\phi_0\cos\phi_0(k\sin^2\phi_0+\ell\cos^2\phi_0)}\\
&\geq\frac{c}{\ell-k}\,\frac{\ell\cot\phi_0+k\tan\phi_0}{k\sin^2\phi_0+\ell\cos^2\phi_0}\\
&>0. \qquad (4.15)
\end{aligned}
$$

We now appeal to an assumption made in (1.101), that the left side of (1.100) has a simple zero at ϕ_0. Thus $T_2(\phi)$ obtained from (4.5) satisfies

$$T_{2,\phi}(\phi_0)<0 \tag{4.16}$$

from (4.12). Using (4.12), (4.15), (4.16), (4.2) and (4.3), we may fix ϕ_2,ϕ_3, satisfying

$$0<\phi_3<\phi_0<\phi_2<\pi/2 \tag{4.17}$$

such that

$$\bar{c}_7\geq T_1(\phi)\geq c_7,\quad \phi_3\leq\phi\leq\phi_2 \tag{4.18}$$

$$T_2(\phi)\geq c_8,\quad 0\geq\phi\leq\phi_3 \tag{4.19}$$

$$\left(\frac{T_2(\phi)}{T_2(\phi)}\right)_\phi\leq -2c_9,\quad \phi_3\leq\phi\leq\phi_2 \tag{4.20}$$

for some positive constants c_7,c_8,c_9, and there exists r_0 such that for all $r\geq r_0$

$$\phi_\xi(r,\phi_2)<0<\phi_\xi(r,\phi_3) \tag{4.21}$$

and

$$-c_{10}r^2\leq r_\xi(r,\phi)\leq -c_{11}r^2,\quad \phi_3\leq\phi\leq\phi_2. \tag{4.22}$$

Ultimately, in the proof of lemma 3.6, we shall fix a value of $\phi_1\in(\phi_3,\phi_0)$ with $\phi_0-\phi_1$ sufficiently small. Then r_0 is chosen sufficiently large, depending on ϕ_1, next ξ_0 sufficiently small, depending on r_0 and ϕ_1, and finally τ assumed sufficiently large.

Making r_0 larger if necessary, using (4.10) and (4.12), it is thus no loss of generality to assume for $r\geq r_0$ that

$$\frac{|R_2(r,\phi)|}{r}\leq\frac{1}{2}T_2(\phi),\quad 0\leq\phi\leq\phi_1 \tag{4.23}$$

from which, using (4.3)

$$\phi_\xi(r,\phi)>0,\quad 0<\phi\leq\phi_1. \tag{4.24}$$

Now from (4.2), (4.3), (4.23), (4.10) and (4.24) again making r_0 larger if necessary,

$$-2r\frac{|T_1(\phi)|}{T_2(\phi)}\leq\frac{r_\xi(r,\phi)}{\phi_\xi(r,\phi)}\leq\begin{cases}\frac{2r|T_1(\phi)|}{T_2(\phi)}, & 0<\phi\leq\phi_3\\ 0, & \phi_3<\phi\leq\phi_1.\end{cases} \tag{4.25}$$

In the interval $0 \le \phi \le \phi_1$, (4.24) allows r, ξ to be described as functions of ϕ, τ, with $r(\xi,\tau) = \hat{r}(\phi(\xi,\tau),\tau)$, $\xi = \hat{\xi}(\phi(\xi,\tau),\tau)$. From (4.2, 4.3)

$$\begin{aligned} \hat{r}_\phi(\hat{r},\phi) &= \frac{r_\xi(\hat{r},\phi)}{\phi_\xi(\hat{r},\phi)} \\ &= -\hat{r}\,\frac{T_1(\phi) - R_1(\hat{r},\phi)/\hat{r}^2}{T_2(\phi) + R_2(\hat{r},\phi)/\hat{r}}, \end{aligned} \tag{4.26}$$

and

$$\hat{\xi}_\phi(\hat{r},\phi) = \frac{1}{\hat{r}T_2(\phi)}\;\frac{1}{\left(1 + \frac{R_2(\hat{r},\phi)}{\hat{r}T_2(\phi)}\right)}, \qquad 0 < \phi \le \phi_1. \tag{4.27}$$

From (4.25) and (4.26), we have an estimate

$$\begin{aligned} &\frac{1}{H(\phi)} \le \frac{\hat{r}(\phi,\tau)}{r(0,\tau)} \le \begin{cases} H(\phi), & 0 < \phi \le \phi_3 \\ H(\phi_3), & \phi_3 < \phi \le \phi_1 \end{cases} \\ &H(\phi) \overset{\text{def}}{=} \exp\left[2\int_0^\phi \frac{|T_1(\phi')|}{T_2(\phi')}d\phi'\right] \end{aligned} \tag{4.28}$$

Using (4.12), (4.23) and (4.28) in (4.27)

$$\begin{aligned} \hat{\xi}(\phi,\tau) &= \int_0^\phi \hat{\xi}_\phi(\hat{r}(\phi',\tau),\phi')d\phi' \\ &= \int_0^\phi \frac{d\phi'}{\hat{r}(\phi',r)T_2(\phi')\,(1 + R_2(\hat{r},\phi')/(\hat{r}T_2(\phi'))} \\ &\le 2\int_0^\phi \frac{d\phi'}{\hat{r}(\phi',\tau)T_2(\phi')} \\ &\le \frac{2}{r(0,\tau)}\int_0^\phi \frac{H(\phi')}{T_2(\phi')}d\phi'. \end{aligned} \tag{4.29}$$

For any fixed ξ_0 and ϕ_1, for sufficiently large τ it is thus no loss of generality using (4.8) and (4.29), to assume

$$\hat{\xi}(\phi_1,\tau) \le \frac{1}{2}\xi_0. \tag{4.30}$$

Next we consider the solution of (4.2, 4.3) in the interval $[\hat{\xi}(\phi_1,\tau),\xi_0]$. As long as $r(\xi,\tau) \ge r_0$, from (4.21) we have

$$\phi_3 \le \phi(\xi,\tau) < \phi_2, \tag{4.31}$$

and thus r_ξ satisfies (4.22).

Introducing the abbreviations

$$\xi_1(\tau) = \hat{\xi}(\phi_1,\tau), r_1(\tau) = r(\xi_1,\tau) = \hat{r}(\phi_1,\tau), \tag{4.32}$$

integrating (4.22) gives

$$\frac{1}{1+c_{10}r_1(\xi_0-\xi_1)} \le \frac{r(\xi_0,\tau)}{r_1} \le \frac{1}{1+c_{11}r_1(\xi_0-\xi_1)} \tag{4.33}$$

so using (4.30)

$$r(\xi_0,\tau) \le \frac{2}{c_{11}\xi_0} \tag{4.34}$$

and as $r_1 \to \infty$ and $\xi_1 \to 0$ as $r \to \infty$

$$\liminf_{r\to\infty} r(\xi_0,\tau) \ge \frac{1}{c_{10}\xi_0}. \tag{4.35}$$

Thus we justify these estimates by choosing

$$\xi_0 < \frac{1}{c_{10}r_0}. \tag{4.36}$$

With both $r(\xi_0,\tau)$ and $\phi(\xi_0,\tau)$ bounded uniformly in τ, from (4.34) and (4.31), respectively, we have $y(\xi_0,\tau)$ bounded uniformly in τ from (4.1). The existence of the limit (3.49) will follow from the uniqueness of the limit of convergent subsequences as $\tau \to \infty$. Suppose this fails, i.e. there exist sequences $\{\tau_m^1\}$, $\{\tau_m^2\}$ increasing without bound as $m \to \infty$, with different limits of $y(\xi_0,\tau)$, i.e.

$$r(\xi_0,\tau_m^1) \to \underline{r}^1, \phi(\xi_0,\tau_m^1) \to \underline{\phi}^1, r(\xi_0,\tau_m^2) \to \underline{r}^2, \phi(\xi_0,\tau_m^2) \to \underline{\phi}^2 \tag{4.37}$$

as $\tau_m^1, \tau_m^2 \to \infty$ with

$$\underline{r}^2 \le \underline{r}^1. \tag{4.38}$$

For $\xi_1 \le \xi \le \xi_0$, r_ξ satisfying (4.22) justifies viewing ϕ, ξ as functions of r, τ,

$$\phi = \tilde{\phi}(r,\tau), \quad \xi = \tilde{\xi}(r,\tau), \quad r_1 \ge r \ge r(\xi_0,\tau) \tag{4.39}$$

with

$$\tilde{\phi}(r_1,\tau) = \phi_1, \quad \tilde{\xi}(r_1,\tau) = \xi_1 \tag{4.40}$$

and recalling that r_1 and ξ_1 depend on τ. From (4.2, 4.3)

$$\tilde{\phi}_r(r,\tilde{\phi}) = \frac{rT_2(\tilde{\phi}) + R_2(r,\tilde{\phi})}{-r^2T_1(\tilde{\phi}) + R_1(r,\tilde{\phi})} \tag{4.41}$$

and

$$\tilde{\xi}_r(r,\tilde{\phi}) = \frac{1}{-r^2T_1(\tilde{\phi}) + R_1(r,\tilde{\phi})}, \quad r(\xi_0,\tau) \le r \le r_1, \tag{4.42}$$

initial values for $\tilde{\phi}$ and $\tilde{\xi}$ obtained from (4.40).

Now from (4.26) and (4.8), $r_1 = \hat{r}(\phi_1,\tau)$ is monotone increasing in τ, and so from (4.40, 4.41), $\tilde{\phi}(r,\tau)$ is monotone increasing in τ for any fixed r, at least for $r < r_1$. But as $r_1 \to \infty$ with τ, this restriction is unnecessary, and thus for any fixed $r > \underline{r}^1$,

$$\tilde{\phi}(r,\tau_m^1), \tilde{\phi}(r,\tau_m^2) \to \bar{\phi}(r) \text{ as } m \to \infty. \tag{4.43}$$

Thus for $(\underline{r}^1, \phi^1) \neq (\underline{r}^2, \phi^2)$ (4.38) must hold with strict inequality. Fix $r_2 > \underline{r}^1$; then using (4.42), (4.43), (4.31), (4.18), (4.10)

$$\begin{aligned}
\tilde{\xi}(r_2, \tau_m^1) - \tilde{\xi}(r_2, \tau_m^2) &= (\xi_0 - \tilde{\xi}(r_2, \tau_m^2)) - (\xi_0 - \tilde{\xi}(r_2, \tau_m^1)) \\
&= -\int_{\underline{r}^2}^{r_2} \tilde{\xi}_r(r, \tilde{\phi}(r, \tau_m^2))dr + \int_{\underline{r}^1}^{r_2} \tilde{\xi}_r(r, \tilde{\phi}(r, \tau_m^1)) \\
&= -\int_{\underline{r}^2}^{\underline{r}^1} \tilde{\xi}_r(r, \tilde{\phi}(r, \tau_m^2))dr + o(1) \quad \text{as} \quad m \to \infty \\
&\geq \frac{1}{\bar{c}_7} \int_{\underline{r}^2}^{\bar{r}^1} \frac{dr}{r^2} + o(1) \quad \text{as} \quad m \to \infty \\
&> \frac{\underline{r}^1 - \underline{r}^2}{2\bar{c}_7 \underline{r}^1 \underline{r}^2}. \qquad (4.44)
\end{aligned}$$

which is positive.

As against this, for any $r_3 > r_2$,

$$\begin{aligned}
\tilde{\xi}(r_2, \tau_m^1) - \tilde{\xi}(r_2, \tau_m^2) &= \xi_1(\tau_m^1) - \xi_1(\tau_m^2) \\
&\quad - \int_{r_2}^{r_1(\tau_m^1)} \tilde{\xi}_r(r, \tilde{\phi}(r, \tau_m^1))dr + \int_{r_2}^{r_1(\tau_m^2)} \tilde{\xi}_r(r, \tilde{\phi}(r, \tau_m^2))dr \\
&= \xi_1(\tau_m^1) - \xi_1(\tau_m^2) - \int_{r_2}^{r_3} (\tilde{\xi}_r(r, \tilde{\phi}(r, \tau_m^1)) - \tilde{\xi}_r(r, \tilde{\phi}(r, \tau_m^2)))dr \\
&\quad - \int_{r_3}^{r_1(\tau_m^1)} \tilde{\xi}_r(r, \tilde{\phi}(r, \tau_m^1))dr + \int_{r_3}^{r_1(\tau_m^2)} \tilde{\xi}_r(r, \tilde{\phi}(r, \tau_m^2))dr. \qquad (4.45)
\end{aligned}$$

In the right side of (4.45), the first terms are $o(1)$ as $r_m^1, r_m^2 \to \infty$ from (4.29) and (4.32); for r_3 fixed, the third term is $o(1)$ as $\tau_m^1, \tau_m^2 \to \infty$ from (4.43), and the last two terms are $O(1/r_3)$ from (4.42), (4.18), (4.31). Thus choosing r_3 sufficiently large and then τ_m^1, τ_m^2 sufficiently large, the right side of (4.45) is made arbitrarily small, contradicting (4.44). Thus the limit (3.49) exists. □

PROOF OF LEMMA 3.5: We rewrite (3.54) with $s = \lambda_+(w_+)$ and $y - w_+$ of the form (3.55), obtaining

$$\begin{aligned}
r_+\zeta_\xi + r_-\omega_\xi &= q(y) - q(w_+) - \lambda_+(w_+)(y - w_+) \\
&= (q_w(w_+) - \lambda_+(w_+))(r_+\zeta + r_-\omega) \\
&\quad + \frac{1}{2} q_{ww}(w_+)(r_+\zeta + r_-\omega)(r_+\zeta + r_-\omega) + R_3(\zeta, \omega) \\
&= -(\lambda_+(w_+) - \lambda_-(w_+))r_-\omega + \frac{\zeta^2}{2} q_{ww}(w_+)r_+r_+ \\
&\quad + \zeta\omega q_{ww}(w_+)r_+r_- + \frac{\omega^2}{2} q_{ww}(w_+)r_-r_- + R_3(\zeta, \omega) \qquad (4.46)
\end{aligned}$$

where the remainder term $R_3(\zeta,\omega)$ satisfies

$$|R_3(\zeta,\omega)| \le c(|\zeta|^3 + |\omega|^3), \tag{4.47}$$

$$|R_{3,\zeta}(\zeta,\omega)|, |R_{3,\omega}(\zeta,\omega)| \le c(\zeta^2 + \omega^2). \tag{4.48}$$

Denote by $\ell_\pm$ the left eigenvectors of $q_w(w_+)$, normalized so that

$$\ell_\pm r_\pm = 1, \quad \ell_\pm r_\mp = 0. \tag{4.49}$$

Multiplying (4.46) by ℓ_-, using (4.49) we obtain (3.57), identifying

$$\begin{aligned} h_2(\zeta,\omega) &= (\frac{1}{2}\ell_- q_{ww}(w_+)r_+r_+)\zeta^2 + (\ell_- q_{ww}(w_+)r_+r_-)\zeta\omega \\ &+ (\frac{1}{2}\ell_- q_{ww}(w_+)r_-r_-)\omega^2 + +\ell_- R_3(\zeta,\omega), \end{aligned} \tag{4.50}$$

and (3.59), (3.62) follow easily from (4.47, 4.48).

Multiplying (4.46) by ℓ_+, using (4.49) we obtain

$$\begin{aligned} \zeta_\xi =&(\frac{1}{2}\ell_+ q_{ww}(w_+)r_+r_+)\zeta^2 + (\ell_+ q_{ww}(w_+)r_+r_-)\zeta\omega \\ &(\frac{1}{2}\ell_+ q_{ww}(w_+)r_-r_-)\omega^2 + \ell_+ R_3(\zeta,\omega) \end{aligned} \tag{4.51}$$

which is (3.56) provided that

$$\frac{1}{2}\ell_+ q_{ww}(w_+)r_+r_+ = -1 \tag{4.52}$$

and identifying the final three right-hand terms in (4.51) as $h_1(\zeta,\omega)$.

The left side of (4.52) is necessarily nonzero by genuine nonlinearity of the system (1.7, 1.8) at the point w_+. Thus (4.52) can always be satisfied by suitable normalization of r_+, and using (4.49).

With $h_1(\zeta,\omega)$ so obtained, (3.58), (3.60) and (3.61) follow directly from (4.47), (4.48).

□

PROOF OF LEMMA 3.6: The integrals (3.64), (3.65) are partitioned into integrals over $(0,\xi_1(\tau))$, within which ϕ and τ are independent variables, with $0 < \phi < \phi_1$, and over the interval $(\xi_1(\tau),\xi_0)$, within which r,τ are the independent variables with $r_1(\tau) > r > r(\xi_0,\tau)$.

Within the first such interval we introduce new dependent variables

$$\nu(\phi,\tau) = \hat{r}(\phi,\tau)/r(0,\tau) \tag{4.53}$$

$$\psi(\phi,\tau) = r(0,\tau)\hat{\xi}(\phi,\tau), \quad 0 \le \phi \le \phi_1. \tag{4.54}$$

From (4.26) and (4.53)

$$\nu_\phi = -\nu\,\frac{T_1(\phi) - \frac{1}{\hat{r}^2}R_1(\hat{r},\phi)}{T_2(\phi) + \frac{1}{\hat{r}}R_2(\hat{r},\phi)}. \tag{4.55}$$

Differentiating (4.55) with respect to τ and using (4.10) we obtain

$$\nu_{\tau\phi} = -\frac{T_1(\phi)}{T_2(\phi)}\nu_\tau + o\Big(|\nu_\tau| + \frac{r_\tau(0,\tau)}{r(0,\tau)}\Big), \quad 0 < \phi < \phi_1 \tag{4.56}$$

as $r(0,\tau) \to \infty$. Ultimately we shall fix

$$r_\tau(0,\tau) = O(r(0,\tau)^{2-\ell}) \tag{4.57}$$

with $\ell > 1$ as appearing in (1.94), so $r_\tau(0,\tau)/r(0,\tau) = o(1)$ as $\tau > \infty$.

As $\nu(0,\tau) = 1$ from (4.53), an initial condition for (4.56) at $\phi = 0$ is

$$\nu_\tau(0,\tau) = 0. \tag{4.58}$$

Then from (4.56) and (4.58)

$$\nu_\tau(\phi,\tau) = o\Big(\frac{r_\tau(0,\tau)}{r(0,\tau)}\Big), \quad 0 < \phi \le \phi_1. \tag{4.59}$$

The estimate (4.59) is uniform with respect to the choice of $\xi_1 \in (\xi_3, \xi_0)$, because $T_1(\phi)$ is positive near ϕ_0, so small positive values of $T_2(\phi)$ make $|\nu_\tau|$ smaller from (4.56).

Similarly from (4.54) and (4.27),

$$\psi_\phi = \frac{1}{\nu T_2(\phi)\left(1 + \frac{R_2(\hat{r},\phi)}{\hat{r}T_2(\phi)}\right)}. \tag{4.60}$$

Differentiating (4.60) with respect to τ, using (4.10), (4.23) and (4.59)

$$\psi_{\tau\phi} = -\frac{\nu_\tau}{\nu^2 T_2(\phi)\Big(1 + \frac{R_2(\hat{r},\phi)}{\hat{r}T_2(\phi)}\Big)} + o\Big(\frac{r_\tau(0,\tau)}{r(0,\tau)}\Big) \tag{4.61}$$

and since $\hat{\xi}(0,\tau) = 0$, $\psi_\tau(0,\tau) = 0$ from (4.54). Integrating (4.61), using (4.59), (4.24), (4.53) and (4.28)

$$|\psi_\tau(\phi)| \le \int_0^\phi (1 + \frac{2H(\phi')^2}{T_2(\phi')})d\phi' + o(\frac{r_\tau(0,\tau)}{r(0,\tau)}), \quad 0 < \phi \le \phi_1. \tag{4.62}$$

Differentiating (4.53) with respect to τ we find

$$\frac{r(0,\tau)}{r_\tau(0,\tau)}\nu_\tau(\phi,\tau) = \frac{\hat{r}_\tau(\phi,\tau)}{r_\tau(0,\tau)} - \frac{\hat{r}(\phi,\tau)}{r(0,\tau)}. \tag{4.63}$$

The left side of (4.63) is o(1) as $\tau \to \infty$ uniformly for $\phi \in (0, \phi_1]$, so using (4.28), for all τ sufficiently large

$$\frac{1}{2H(\phi)} \le \frac{\hat{r}_\tau(\phi,\tau)}{r_\tau(0,\tau)} \le 2H(\phi_3), \quad 0 < \phi \le \phi_1. \tag{4.64}$$

Differentiating (4.54) with respect to τ, we use (4.29), (4.54) and (4.62) to obtain

$$|\hat{\xi}_\tau(\phi,\tau)| \le r_\tau(0,\tau)G_1(\phi,\tau)/r(0,\tau)^2, \tag{4.65}$$

$$G_1(\phi,\tau) = 2\int_0^\phi \frac{H(\phi')}{T_2(\phi')}d\phi' + o(1)\int_0^\phi (1 + \frac{2H(\phi')^2}{T_2(\phi')}d\phi', \quad 0 < \phi \le \phi_1. \tag{4.66}$$

In the region $\xi_1(\tau) \le \xi \le \xi_0$, corresponding to $r_1(\tau) \ge r \ge r(\xi_0,\tau)$, r satisfies (4.22) and $\tilde{\phi}(r,\tau) = \phi(\tilde{\xi}(r,\tau),\tau)$ satisfies (4.31). Thus $T_1(\phi)$ is bounded away from zero from (4.18). Differentiating (4.41) with respect to τ, using (4.10) and (4.20), we find

$$\tilde{\phi}_{\tau r} = -\frac{1}{r}\Big(\big(\frac{T_2(\tilde{\phi})}{T_1(\tilde{\phi})}\big)_\phi + o(1)\Big)\tilde{\phi}_\tau \qquad \text{as } r \to \infty$$

$$\ge \frac{c_9}{r}\tilde{\phi}_\tau \tag{4.67}$$

making r_0 larger as necessary, without loss of generality.

Differentiating (4.42) with respect to τ and using (4.10) we have a constant c_{14} such that

$$|\tilde{\xi}_{\tau r}| \leq c_{14} \frac{|\tilde{\phi}_\tau|}{r^2}. \tag{4.68}$$

To obtain values for $\tilde{\phi}_\tau(r_1,\tau)$ and $\tilde{\xi}_\tau(r_1,\tau)$ we differentiate (4.32) and (4.40) with respect to τ, obtaining

$$\tilde{\phi}_\tau(r_1,\tau) = -\tilde{\phi}_r(r_1,\tau)\hat{r}_\tau(\phi_1,\tau) \tag{4.69}$$

and

$$\tilde{\xi}_\tau(r_1,\tau) = \hat{\xi}_\tau(\phi_1,\tau) - \tilde{\xi}_r(r_1,\tau)\hat{r}_\tau(\phi_1,\tau). \tag{4.70}$$

Setting $\phi = \phi_1$ in (4.53) and differentiating with respect to τ, using (4.59) we have

$$\hat{r}_\tau(\phi_1,\tau) = r_\tau(0,\tau)\left(\frac{r_1(\tau)}{r(0,\tau)} + o(1)\right) \qquad \text{as } \tau \to \infty, \text{uniformly in } \phi_1. \tag{4.71}$$

We use (4.41) and (4.71) in (4.69) to obtain

$$\frac{\tilde{\phi}_\tau(r_1,\tau)}{r_\tau(0,\tau)} = -\frac{T_2(\phi_1)}{r_1 T_1(\phi_1)} \frac{\left(1+\frac{R_2(r_1,\phi_1)}{r_1 T_2(\phi_1)}\right)}{\left(1-\frac{R_1(r_1,\phi_1)}{r_1^2 T_1(\phi_1)}\right)} \left(\frac{r_1}{r(0,\tau)} + o(1)\right) \tag{4.72}$$

and using (4.10), (4.18), (4.23), and (4.28) in (4.72), we get

$$|\hat{\phi}_\tau(r_1,\tau)| \leq \frac{2T_2(\phi_1)}{c_7} \frac{r_\tau(0,\tau)}{r(0,\tau)} \left(1 + o(1)H(\phi_1)\right), \quad \text{as } \tau \to \infty. \tag{4.73}$$

Setting $\phi = \phi_1$ in (4.65)

$$|\hat{\xi}_\tau(\phi_1,\tau)| \leq \frac{r_\tau(0,\tau)}{r(0,\tau)^2} G_1(\phi_1,\tau). \tag{4.74}$$

Using (4.74), (4.42), (4.71), (4.18) and (4.28) in (4.70) we get

$$|\tilde{\xi}_\tau| \leq \frac{r_\tau(0,\tau)}{r(0,\tau)^2} G_2(\phi_1,\tau) \tag{4.75}$$

with

$$G_2(\phi_1,\tau) = \frac{2}{c_7}\left(H(\phi_1) + o(1)H(\phi_1)^2\right) + G_1(\phi_1,\tau). \tag{4.76}$$

Thus from (4.67) and (4.73)

$$|\tilde{\phi}_\tau(r,\tau)| \leq \frac{2}{c_7} T_2(\phi_2) \frac{r_\tau(0,\tau)}{r(0,\tau)} (1 + o(1)H(\phi_1)) \left(\frac{r}{r_1}\right)^{c_9}, \quad r_0 \leq r \leq r_1(\tau). \tag{4.77}$$

From (4.68) and (4.75), there is a constant c_{15} such that

$$|\tilde{\xi}_\tau(r,\tau)| \leq \frac{r_\tau(0,\tau)}{r(0,\tau)^2} G_2(\phi_1,\tau) + c_{15}(\phi_\tau(r,\tau)|/r_1(\tau). \tag{4.78}$$

To get the estimate (3.65) we use

$$\int_0^{\xi_0} |y_\tau(\xi,\tau)| d\xi \leq \int_0^{\xi_0} \left(\ell r(\xi,\tau)^{\ell-1} |r_\tau(\xi,\tau)| + r(\xi,\tau)^\ell |\phi_\tau(\xi,\tau)|\right) d\xi. \tag{4.79}$$

The integral (4.79) is partitioned at $\xi_1(\tau)$, using the elementary relations

$$r_\tau(\hat{\xi},\tau) = \hat{r}_\tau(\phi,\tau) + \hat{r}_\phi(\hat{r},\phi)\phi_\tau(\hat{\xi},\tau) \tag{4.80}$$

$$\phi_\tau(\hat{\xi},\tau) = -\phi_\xi(\hat{r},\phi)\hat{\xi}_\tau(\phi,\tau) \tag{4.81}$$

in the interval $(0,\xi_1(\tau))$ and

$$r_\tau(\tilde{\xi},\tau) = -\tilde{\xi}_\tau(r,\tau)/\tilde{\xi}_r(r,\tilde{\phi}), \tag{4.82}$$

$$\phi_\tau(\tilde{\xi},\tau) = \tilde{\phi}_\tau(r,\tau) + \tilde{\phi}_r(r,\tau)r_\tau(\tilde{\xi},\tau) \tag{4.83}$$

in the interval $(\xi_1(\tau),\xi_0)$.

Using (4.80-4.83) and changing the variable of integration in (4.79) we obtain

$$\int_0^{\xi_0} |y_\tau(\xi,\tau)|d\xi \le \int_0^{\xi_1(\tau)} \hat{r}(\phi,\tau)^{\ell-1}\left[\frac{\ell|\hat{r}_\tau(\phi,\tau)|}{\phi_\xi(\hat{r},\phi)} + (\ell|\hat{r}_\phi(\hat{r},\phi)| + \hat{r})|\hat{\xi}_\tau(\phi,\tau)|\right]d\phi \tag{4.84}$$
$$+ \int_{r(\xi_0,\tau)}^{r_1(\tau)} r^{\ell-1}\left[r|\tilde{\phi}_\tau(r,\tau)\tilde{\xi}_r(r,\tilde{\phi})| + |\tilde{\xi}_\tau(r,\tau)|(\ell + r|\tilde{\phi}_r(r,\tilde{\phi})|)\right] dr.$$

In the first right-hand term of (4.84), $\hat{r}$ is estimated using (4.28); $\hat{r}_\tau$ obtained from (4.64); ϕ_ξ from (4.3); $\hat{r}_\phi$ from (4.26); and $\hat{\xi}_\tau$ from (4.65). In the second term, $\tilde{\phi}_\tau$ is obtained from (4.77); $\tilde{\xi}_r$ from (4.42); $\tilde{\xi}_\tau$ from (4.78); and $\tilde{\phi}_r$ from 14.41). We also employ (4.18) and (4.23) in simplifying the resulting estimate, of the form

$$\int_0^{\xi_0} |y_\tau(\xi,\tau)|d\xi \le r(0,\tau)^{\ell-2} r_\tau(0,\tau)G_3(\phi_1,\tau) \tag{4.85}$$

where $G_3(\phi_1,\tau)$ is uniformly bounded with respect to τ for any fixed $\phi_1 < \phi_0$. Thus with $r_\tau(0,\tau)$ chosen satisfying (4.57), we will satisfy (3.65) using (4.85), (4.57).

This choice for $r_\tau(0,\tau)$ also suffices to satisfy (3.66). Indeed, with $r_\tau(0,\tau)$ satisfying (4.57), from (4.77) and (4.78)

$$\lim_{\tau\to\infty} \tilde{\phi}_\tau(r,\tau), \lim_{\tau\to\infty} \tilde{\xi}_\tau(r,\tau) = 0 \tag{4.86}$$

uniformly with respect to r in any bounded interval. As $r(\xi_0,\tau)$ is uniformly bounded with respect to τ, by appeal to lemma 3.4, (3.64) follows from

$$|y_\tau(\xi_0,\tau)| \le r^{\ell-1}|r_\xi(r,\phi)|\left[r|\tilde{\phi}_\tau(r,\tau)\tilde{\xi}_r(r,\phi)| + |\tilde{\xi}_\tau(r,\tau)|(\ell + r|\tilde{\phi}_r(r,\phi)|\right), \tag{4.87}$$

obtained in the same way as (4.84), with $r = r(\xi_0,\tau), \phi = \phi(\xi_0,\tau)$.

The delicate step in the proof is thus obtaining (3.64). We employ the same changes of independent variable,

$$\int_0^{\xi_0} y_\tau(\xi,\tau)d\xi = \frac{d}{d\tau}\Big(\int_0^{\xi_0} y(r(\xi,\tau),\phi(\xi,\tau))d\xi\Big)$$

$$= \frac{d}{d\tau}\Big(\int_0^{\xi_1(\tau)} y(r(\xi,\tau),\phi(\xi,\tau))d\xi + \int_{\xi_1(\tau)}^{\xi_0} y(r(\xi,\tau),\phi(\xi,\tau))d\xi\Big)$$

$$= \frac{d}{d\tau}\Big(\int_0^{\phi_1} \frac{y(\hat{r}(\phi,\tau),\phi)}{\phi_\xi(\hat{r}(\phi,\tau),\phi)}d\phi - \int_{r(\xi_0,\tau)}^{r_1(\tau)} \frac{y(r,\tilde{\phi}(r,\tau))}{r_\xi(r,\tilde{\phi}(r,\tau))}dr\Big)$$

$$= \int_0^{\phi_1}\Big((\frac{y}{\phi_\xi})_r(\hat{r}(\phi,\tau),\tau)\Big)\hat{r}_\tau(\phi,\tau)d\phi - \frac{y(r_1(\tau),\phi_1)}{r_\xi(r_1(\tau),\phi_1)}r_{1,\tau}(\tau)$$

$$+\frac{y(r(\xi_0,\tau),\phi(\xi_0,\tau))}{r_\xi(r(\xi_0,\tau),\phi(\xi_0,\tau))}r_\tau(\xi_0,\tau) - \int_{r(\xi_0,\tau)}^{r_1(\tau)}\Big((\frac{y}{r_\xi})_\phi(r,\tilde{\phi}(r,\tau))\tilde{\phi}_\tau(r,\tau)dr. \tag{4.88}$$

For each of the two components of y, we have $y(r,\phi)$ from (4.1). For each component of y, we estimate each term in the right side of (4.88), using the available equations and estimates: ϕ_ξ in (4.3); r_ξ in (4.2); $\hat{r}$ in (4.28); $\tilde{\phi}$ in (4.31); $\hat{r}_\tau$ in (4.64); $r_{1,\tau}$ in (4.71); $r(\xi_0,\tau)$ in (4.34); $\phi(\xi_0,\tau)$ from (4.31); $r_\tau(\xi_0,\tau)$ from (4.82), with $\tilde{\xi}=\xi_0$; and $\tilde{\phi}_\tau$ from (4.77). With these substitutions, the right side of (4.88) becomes homogeneous of degree one in $r_\tau(0,\tau)$.

With $y=y_1=-r^k\sin\phi$, $k\le 1$, each of the four terms in the right side of (4.88) is of $o(1)$ as $\tau\to\infty$ for any fixed $\phi_1<\phi_0$ and $r_\tau(0,\tau)$ satisfying (4.57). Thus the first component of (3.64) is satisfied.

With $y=y_2=r^\ell\cos\phi$, $\ell>1$, the first right-hand term in (4.88) is bounded above and below by positive constants, the lower bound following from the lower bounds in (4.64) and (4.57). Furthermore, these constants are uniform with respect to ϕ_1. The second term is positive and bounded from above, but not uniformly with respect to ϕ_1. The third term is $o(1)$.

The fourth term is majorized by $T_2(\phi_1)r_1(\tau)/r(o,\tau)$. As $r_1(\tau)/r(0,\tau)$ is bounded uniformly in τ, setting $\phi=\phi_1$ in (4.28), we choose ϕ_1 close to ϕ_0 to make $T_2(\phi_1)$ small, using (4.11). In particular we want the fourth term in (4.88) strictly smaller than the first, so that the second component of the right side of (4.88) is an increasing function of $r_\tau(0,\tau)$. Thus $r_\tau(0,\tau)$ can be chosen satisfying (4.57) so that the second component of (3.64) is also satisfied.

□

Proof of lemma 3.7: By appeal to lemma 3.5, it suffice to consider a one-parameter family of solutions of (3.56, 3.57) in the interval $2<\xi<\infty$, with h_1,h_2 satisfying (3.58-3.62), $\kappa>0$ and c_5 given. From (3.53) and (3.63), we have initial data at $\xi=2$ satisfying

$$\zeta(2,\tau)\to c_6,\ \omega(2,\tau)\to 0 \ \text{ as } \tau\to\infty \tag{4.89}$$

where $c_6>0$ is chosen sufficiently small below.

Applying lemma 3.6, using (3.66) in (3.55), we have in addition

$$\zeta_\tau(2,\tau), \omega_\tau(2,\tau) \to 0 \text{ as } \tau \to \infty. \tag{4.90}$$

Using (3.55), the desired results (3.14), (3.69) will follow, respectively, from

$$\zeta(\xi,\tau), \omega(\xi,\tau) \to 0 \text{ as } \xi \to \infty \tag{4.91}$$

$$\int_2^\infty (|\omega_\tau(\xi,\tau)| + |\zeta_\tau(\xi,\tau)|)d\xi \to 0 \text{ as } \tau \to \infty \tag{4.92}$$

$$\omega_\tau(\cdot,\tau), \zeta_\tau(\cdot,\tau) \in L_1(2,\infty) \text{ pointwise in } \tau. \tag{4.93}$$

Using (4.90), to obtain (4.92-4.93) it suffices to show that

$$|\omega_\tau(\xi,\tau)| + |\zeta_\tau(\xi,\tau)| \le \frac{c}{1+c_6^2\xi^2}(|\omega_\tau(2,\tau)| + |\zeta_\tau(2,\tau)|) \tag{4.94}$$

where here and below, unsubscripted c is a generic constant, independent of c_6 and τ.

The result (4.94) is curious in that $|\zeta_\tau|$ decays faster than $|\zeta|$ for large ξ; in particular $\zeta(\cdot,\tau) \notin L_1(2,\infty)$.

Denote by

$$\sigma(\xi,\tau) = \frac{\zeta(\xi,\tau)}{1-(\xi+\frac{1}{2c_6})\zeta(\xi,\tau)} \tag{4.95}$$

and

$$b(\xi,\tau) = 1 + (\xi + \frac{1}{2c_6})\sigma(\xi,\tau) \tag{4.96}$$

for $2 \le \xi \le \xi_4$, where ξ_4 is small enough, possibly depending on τ, so that for $2 \le \xi < \xi_4$

$$0 < \sigma(\xi,\tau) < \infty, |\omega(\xi,\tau)| \le \frac{\kappa}{2c_5}. \tag{4.97}$$

Within this interval, from (4.95), (4.96)

$$\begin{aligned} 0 < \zeta(\xi,\tau) &= \frac{\sigma(\xi,\tau)}{b(\xi,\tau)} \\ &< \frac{1}{\xi+\frac{1}{2c_6}} \end{aligned}, \tag{4.98}$$

and using (4.95), (4.96), we rewrite (3.56)

$$\sigma_\xi = \frac{1}{b^2}h_1(\frac{\sigma}{b},\omega). \tag{4.99}$$

Using (4.97), (4.98) in (3.59)

$$|h_2(\zeta,\omega)| \le \frac{\kappa}{2}|\omega| + \frac{c_5}{(\xi+\frac{1}{2c_6})^2} \tag{4.100}$$

so (3.57) gives an estimate

$$|\omega|_\zeta \le -\frac{\kappa}{2}|\omega| + c_5/(\xi + \frac{1}{2c_6})^2 \tag{4.101}$$

which is readily integrated to obtain

$$\begin{aligned}\left|\omega(\xi,\tau)-\omega(2,\tau)e^{-\frac{\kappa}{2}(\xi-2)}\right| &\le c\int_2^{\xi} e^{-\frac{\kappa}{2}(\xi-\xi')}\frac{d\xi'}{(\xi'+\frac{1}{2c_6})^2}\\ &\le \frac{c}{(\xi+\frac{1}{2c_6})^2}+c\int_2^{\xi} e^{-\frac{\kappa}{2}(\xi-\xi')}\frac{d\xi'}{(\xi'+\frac{1}{2c_6})^3}\\ &\le \frac{c}{(\xi+\frac{1}{2c_6})^2}+cc_6^3\int_2^{\xi/2} e^{-\frac{\kappa}{2}(\xi-\xi')}d\xi'+c\int_{\xi/2}^{\xi}\frac{d\xi'}{(\xi'+\frac{1}{2c_6})^3}\\ &\le \frac{c}{(\xi+\frac{1}{2c_6})^2}\quad,\end{aligned}\tag{4.102}$$

employing a partial integration in the first step.

Now using (4.98) and (4.102) in (3.58), we have

$$|h_1(\zeta,\omega)|\le c/(\xi+\frac{1}{2c_6})^3;\tag{4.103}$$

using (4.103) in (4.99), integrating with respect to ξ we find

$$|\sigma(\xi,\tau)-\sigma(2,\tau)|\le cc_6^2.\tag{4.104}$$

From (4.95) and (4.89),

$$\sigma(2,\tau)=2c_6/(1-4c_6)+o(1)\quad\text{as}\quad\tau\to\infty,\tag{4.105}$$

so taking c_6 sufficiently small, we have

$$c_6<\sigma(\xi,\tau)<3c_6,\quad 2\le\xi<\xi_4.\tag{4.106}$$

Now from (4.97), (4.102), (4.106), making c_6 smaller if necessary, we can have $\xi_4=\infty$, so (4.102), (4.106) and (4.98) hold for all ξ,τ, and (4.91) is proved.

From (4.96) and (4.106),

$$\frac{1}{2}(3+c_6\xi)\le b(\xi,\tau)\le\frac{5}{2}+3c_6\xi.\tag{4.107}$$

Differentiating (4.95) with respect to τ and using (4.96) we obtain

$$|\zeta_\tau(\xi,\tau)|\le\frac{|\sigma_\tau(\xi,\tau)|}{b(\xi,\tau)^2}\quad;\tag{4.108}$$

let ξ_5, possibly depending on τ, be such that

$$|\sigma_\tau(\xi,\tau)|\le 2|\sigma_\tau(2,\tau)|+|\omega_\tau(2,\tau)|,\quad 2<\xi<\xi_5.\tag{4.109}$$

Differentiating (3.57) with respect to τ and using (3.62) we have, within the interval $2<\xi<\xi_5$,

$$\begin{aligned}|\omega_\tau|_\xi &\le -\kappa|\omega_\tau|+c_5(|\zeta|+|\omega|)\ (|\zeta_\tau|+|\omega_\tau|)\\ &\le -\kappa|\omega_\tau|+(\frac{cc_6}{1+2c_6\xi})\ (\frac{|\sigma_\tau|}{b^2}+|\omega_\tau|)\end{aligned}\tag{4.110}$$

using (4.98), (4.102) and (4.108). Making c_6 smaller if necessary and using (4.107), (4.109), we obtain from (4.110)

$$|\omega_\tau|_\xi \leq -\frac{\kappa}{2}|\omega_\tau| + \frac{cc_6|\sigma_\tau(2,\tau)|}{b^3}. \tag{4.111}$$

By the same technique used to obtain (4.102) from (4.101), we obtain from (4.111)

$$|\omega_\tau(\xi,\tau)| \leq |\omega_\tau(2,\tau)|e^{-\frac{\kappa}{2}(\xi-2)} + \frac{c\,c_6|\sigma_\tau(2,\tau)|}{b(\xi,\tau)^3} \quad . \tag{4.112}$$

Differentiating (4.99) with respect to τ, using (3.58), (3.60), (3.61), (4.96) and (4.108)

$$\begin{aligned}|\sigma_\tau|_\xi &\leq 2\frac{|h_1|}{b^3}b_\tau + \frac{1}{b^2}|h_{1,\zeta}|\,|\zeta_\tau| + \frac{1}{b^2}|h_{1,\omega}|\,|\omega_\tau| \\ &\leq 2c_5(|\zeta|^3 + |\zeta\omega| + |\omega|^2)\frac{(\xi + \frac{1}{2c_6})}{b^3}|\sigma_\tau| \\ &\quad + \frac{c_5}{b^2}(|\zeta|^2 + |\omega|)\frac{|\sigma_\tau|}{b^2} + \frac{c_5}{b^2}(|\zeta| + |\omega|)|\omega_\tau|.\end{aligned} \tag{4.113}$$

Using (4.98) and (4.102) in (4.113)

$$|\sigma_\tau|_\xi \leq c\frac{|\sigma_\tau|}{b^3(\xi + \frac{1}{2c_6})^2} + \frac{c}{b^2(\xi + \frac{1}{2c_6})}|\omega_\tau|\,. \tag{4.114}$$

Now using (4.109) and (4.112) in (4.114)

$$\begin{aligned}|\sigma_\tau|_\xi &\leq c|\sigma_\tau(2,\tau)|\Big(\frac{1}{b^3(\xi + \frac{1}{2c_6})^2} + \frac{c_6}{b^5(\xi + \frac{1}{2c_6})}\Big) \\ &\quad + \frac{cc_6|\omega_\tau(2,\tau)|}{b^2}e^{-\frac{\kappa}{2}(\xi-2)}.\end{aligned} \tag{4.115}$$

Integrating (4.115) •vith respect to ξ, we have

$$|\sigma_\tau(\xi,\tau) - \sigma_\tau(2,\tau)| \leq cc_6(|\sigma_\tau(2,\tau)| + |\omega_\tau(2,\tau)|) \tag{4.116}$$

so making c_6 smaller, if necessary, (4.109) holds for all ξ and we may take $\xi_5 = \infty$. Now (4.94) follows from (4.108), (4.109) and (4.112).

□

5. Whence they came, where they went

The origin and fate of singular shocks is discussed here, in the context of solution of the Cauchy problem for systems of structure described in sections 1–3 above. For simplicity, we tacitly assume here smooth initial data of compact support, but large oscillation.

For the model problem (2.17, 2.18), such solution has been obtained by application of a front-tracking algorithm [KLS]. In general distribution solutions, including a finite number of singular shocks, are obtained. The appearance of singular shocks is supported by numerical computations [SS], perhaps surprisingly well.

Features of the argument given for the model problem can be generalized, providing a point of departure for the present study. The given proof of convergence of a sequence of front-tracking algorithms is not particular to the model problem, but does depend on its exceptional symmetry group. We shall illustrate this in section 6 below in solving the closely related, but nowhere hyperbolic system (2.24, 2.25).

Our attention here is focused on the composition of the wave sets $\mathcal{A}_\delta, \delta > 0$, associated with front-tracking approximations of solutions including singular shocks. The composition of $\mathcal{A}_\delta$ presumably determines the local features of solutions obtained by front-tracking. In this context, the possibility of including measures of the form $(2.7)_1$ in $\mathcal{A}_\delta$ was discussed in section 2 of chapter 1. Existence of the corresponding front-tracking approximation in general requires that the set $\mathcal{A}_\delta$ satisfy a completeness condition.

DEFINITION. **A wave set $\mathcal{A}_\delta$ is complete for a given system of conservation laws if the initial value problems arising from the interaction of any finite number of elements of $\mathcal{A}_\delta$ are solvable within the set $\mathcal{A}_\delta$.**

We shall not demand uniqueness here. Given multiple solutions, one presumably simply chooses the one minimizing the variation of the solution, as measured by whatever adopted metric.

In the context of obtaining ordinary weak solutions but including strong shocks, the elements of the set $\mathcal{A}_\delta$ are simple discontinuities, of the form (3.1), corresponding to entropy shocks (or contact discontinuities) of whatever strength, and entropy-violating shocks (or suitable approximations thereof) of strength not exceeding δ.

Completeness of $\mathcal{A}_\delta$ then follows from the solvability of classical Riemann problems, with initial data of the form

$$w(x,0) = \begin{cases} w_\ell, & x < 0\,, \\ w_r, & x > 0 \end{cases} \qquad w_\ell,\ w_r \in D \tag{5.1}$$

in the class of self-similar solutions. However, given a hyperbolic system with bounded Hugoniot loci, quite generally the interaction of two sufficiently strong entropy shocks determines a classically unsolvable Riemann problem. These problems can be solved by including some measures of the form $(2.7)_1$ in $\mathcal{A}_\delta$, with nonzero singular mass in some predetermined set $\mathcal{M} \subset \mathbb{R}^n$. Completeness of $\mathcal{A}_\delta$ now requires, in addition, the solution within $\mathcal{A}_\delta$ of a class of "extended Riemann problems", with initial data of the form

$$w(x,0) = \begin{cases} w_\ell, & x < 0\,, \\ w_r, & x > 0 \end{cases} + M_0\delta(x), \quad w_\ell, w_r \in D, \quad M_0 \in \mathcal{M}, \tag{5.2}$$

such problems arising from the interactions of the measures in $\mathcal{A}_\delta$ with whatever other elements.

For the model problem, the set $\mathcal{A}_\delta$ is selected ad hoc in [KLS] and shown to be complete. Indeed, both (5.1) and (5.2) are solved uniquely within $\mathcal{A}_\delta$ for any $w_\ell, w_r \in \mathbb{R}^2$ and any $M_0 \in \mathcal{M}$. The uniqueness results are exceptional, but results on the structure of the set $\mathcal{A}_\delta$ can be generalized.

We consider pairs of conservation laws of the form (1.7), (1.8), strictly hyperbolic and genuinely nonlinear in a simply connected region $H \subseteq D \subseteq \mathbb{R}^2$. The characteristic speeds $\lambda_\pm$, corresponding right eigenvectors $r_\pm$, and Riemann invariants $\nu_\pm$ are smooth functions of w within H; indeed we assume that the mapping of $w \to \begin{pmatrix}\nu_+\\ \nu_-\end{pmatrix}$ is an isomorphism of $H \to \begin{pmatrix}\nu_+\\ \nu_-\end{pmatrix}(H)$. Then throughout H by convention

$$\lambda_- < \lambda_+, \tag{5.3}$$

$$\nabla\nu_\pm \cdot r_\pm = 0, \tag{5.4}$$

$$r_\pm = \frac{\partial w}{\partial \nu_\mp} \tag{5.5}$$

and

$$\frac{\partial \lambda_\pm}{\partial \nu_\mp} > 0. \tag{5.6}$$

We assume systems satisfying the condition (1.86) and assumptions of theorem 3.3, specifically (1.92–1.94), (1.99–1.101), (3.39) and (3.40). In addition, we make an assumption somewhat stronger than implied by strict hyperbolicity and (1.99), that for all $w_0 \in H$,

$$\frac{\partial f}{\partial v}(w_0) < 0. \tag{5.7}$$

These assumptions are satisfied, for example, by the model problem (2.17, 2.18), and the systems (2.19, 2.20) with the limitation (2.4). We do not need assumptions here on the entropy density or symmetry group of a given system.

Two implications of the assumption (5.7) follow. Denote by

$$\begin{aligned} &\Lambda_0(w_0)(\text{resp.}\Lambda_+(w_0), \Lambda_-(w_0)) \\ &= \{w \in H \mid u = u_0(\text{ resp. } u > u_0, u < u_0)\} \end{aligned} \tag{5.8}$$

for all $w_0 \in H$. From (1.9), (5.7) and (5.8), it follows that

$$\Gamma(w_0) \cap \Lambda_0(w_0) = \{w_0\} \tag{5.9}$$

and we designate two subsets of $\Gamma(w_0)$

$$\Gamma_\pm(w_0) = \Gamma(w_0) \cap (\Lambda_\pm(w_0) \cup \Lambda_0(w_0)) \tag{5.10}$$

intersecting only at w_0.

Next using (5.5), the first component of the characteristic equation for a system (1.7) is

$$(f_u - \lambda_\pm)\frac{\partial u}{\partial \nu_\mp} + f_v \frac{\partial v}{\partial \nu_\mp} = 0. \tag{5.11}$$

From (5.7) and (5.11), since $r_\pm$ never vanish, it follows that each of $\partial u/\partial \nu_+$, $\partial u/\partial \nu_-$ is nonvanishing within H. We shall assume slightly more, that throughout H

$$\frac{\partial u}{\partial \nu_\pm} > 0\,. \tag{5.12}$$

This assumption is consistent with the above cited examples and is explained further below.

Our first objective is the "large data" solution of the Riemann problem for such systems, in the class of entropy shocks and rarefaction waves. Such solution obviously depends on the structure of the Hugoniot locus of a given point in H, which is only roughly determined by the above assumptions. In particular, the condition (1.99) implies that $\Gamma(w_0)$ is bounded for any $w_0 \in$H.

The simplest form for $\Gamma(w_0)$, consistent with the above assumptions and the cited examples, is obtained with each of $\Gamma_+(w_0), \Gamma_-(w_0)$ a simple closed curve, each containing only two points at which the shock speed becomes characteristic, as shown in figures 5.1 and 5.2.

These points are designated

$$P_1(w_0) \in \Gamma_-(w_0), \quad s(P_1(w_0), w_0) = \lambda_-(w_0) \tag{5.13}$$

$$P_2(w_0) \in \Gamma_-(w_0), \quad s(P_2(w_0), w_0) = \lambda_+(P_2(w_0)) \tag{5.14}$$

$$P_1'(w_0) \in \Gamma_+(w_0), \quad s(P_1'(w_0), w_0) = \lambda_+(w_0) \tag{5.15}$$

$$P_2'(w_0) \in \Gamma_+(w_0), \quad s(P_2'(w_0), w_0) = \lambda_-(P_2'(w_0)) \tag{5.16}$$

We adopt a standard entropy condition based on comparison of shock speeds and characteristic speeds [L1]. A point w_- is connected on the left to a point $w_+ \in \Gamma(w_-)$ on the right by an entropy j-shock if

$$\lambda_{j-1} w_-, \lambda_j(w_+) \leq s(w_-, w_+) \leq \lambda_j(w_-), \ \lambda_{j+1}(w_+) \tag{5.17}$$

for some characteristic family j; by convention in (5.17), $\lambda_0 = -\infty$ and $\lambda_{n+1} = \infty$.

For $w_0 \in H$, denote by $S_1(w_0), S_2(w_0)$ the sets of entropy 1- and 2-shocks connecting w_0 on the left, and $S_1'(w_0)$, $S_2'(w_0)$ similarly connecting w_0 on the right. Each of these is a half-open segment within $\Gamma(w_0)$; moving counterclockwise on $\Gamma(w_0)$ in figure 5.1, we identify

$$S_2(w_0) = (w_0, P_1(w_0)], \tag{5.18}$$

$$S_1(w_0) = [P_2(w_0), w_0). \tag{5.19}$$

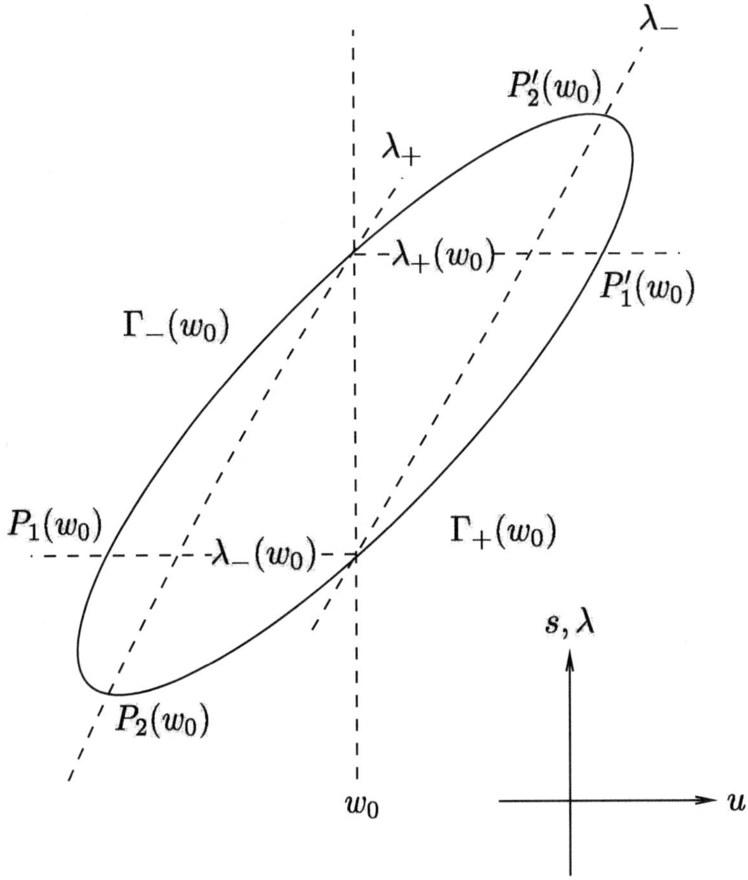

Figure 3.5.1: Shock speed and exceptional points on $\Gamma(w_0)$

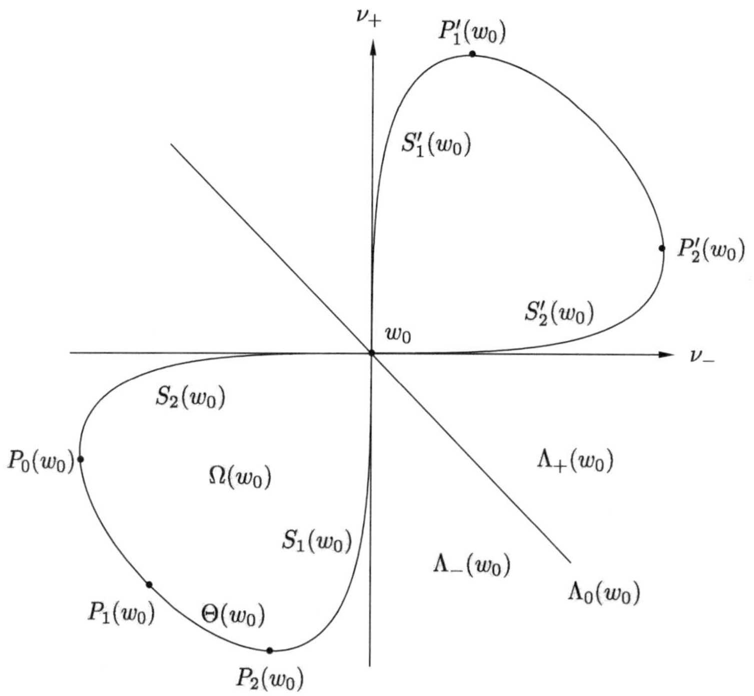

Figure 3.5.2: Hugoniot locus using Riemann invariants coordinates

Moving counterclockwise on $\Gamma_+(w_0)$, we obtain similarly

(5.20) $$S_1'(w_0) = (w_0, P_1'(w_0)]$$

(5.21) $$S_2'(w_0) = [P_2'(w_0), w_0).$$

It follows from (5.14), (5.16) that $\Gamma(w_0)$ continues in the ν_--direction at $P_2(w_0)$ and in the ν_+-direction at $P_2'(w_0)$. Within $\Gamma(w_0)$, typically the minimum of ν_+ is achieved at $P_2(w_0)$ and the maximum of ν_- at $P_2'(w_0)$.

In general $P_1(w_0), P_1'(w_0)$ are not such local extremum points, and this will be important in the discussion of uniqueness. In particular, we denote by $P_0(w_0)$ the point(s) achieving the minimum of ν_- within $S_2(w_0)$.

The open interior of $\Gamma_-(w_0)$ is denoted by $\Omega(w_0)$.

We comment briefly on the assumption (5.12). If $\partial u/\partial \nu_+$ and $\partial u/\partial \nu_-$ were of opposite sign, the structure of $\Gamma(w_0)$ would necessarily be more complicated. No such examples are known to this author. Making $\partial u/\partial \nu_\pm$ both negative simply reverses the direction of the singular mass in the singular shocks, otherwise leading to entirely similar conclusions.

By inspection of figure 5.1, the remaining open segment of $\Gamma_-(w_0)$

(5.22) $$\Theta(w_0) = (P_1(w_0), P_2(w_0))$$

moving counterclockwise, corresponds to strictly overcompressive shocks, satisfying

(5.23) $$\lambda_+(w) < s(w, w_0) < \lambda_-(w_0), \quad w \in \Theta(w_0).$$

The weak overcompressibility condition (3.43) is satisfied for $w \in \bar{\Theta}(w_0)$, the closure of $\Theta(w_0)$.

For a given point $w_0 \in H$, the points which can be connected on the right to w_0 on the left by a 1- or 2- rarefaction wave are given immediately from (5.5), (5.6),

$$R_1(w_0) = \{w \in H | \nu_-(w) = \nu_-(w_0),\ \nu_+(w) > \nu_+(w_0)\} \tag{5.24}$$

$$R_2(w_0) = \{w \in H | \nu_+(w) = \nu_+(w_0),\ \nu_-(w) > \nu_-(w_0)\} \tag{5.25}$$

Now given w_ℓ, $w_r \in H$, we seek a classical similarity solution of the Riemann problem (5.1), with w_ℓ connected on the left to some $w_m \in H$ on the right by an entropy 1-shock or 1-rarefaction (or $w_\ell = w_m$), and w_m connected on the left to w_r on the right by an entropy 2-shock or 2-rarefaction (or $w_m = w_r$). Among the standard four generic cases, the "shock-shock" case requires special treatment because of the form of the Hugoniot locus. In this case we have

$$w_m \in S_1(w_\ell),\ \ w_r \in S_2(w_m), \tag{5.26}$$

with S_1, S_2 given explicitly in (5.19), (5.18); single-valuedness of the solution requires

$$s(w_\ell, w_m) \le s(w_m, w_r). \tag{5.27}$$

LEMMA 5.1. *Assume w_ℓ, w_m, w_r satisfy (5.26) and (5.27) with equality. Then w_r is connected to w_ℓ by a strictly overcompressive shock,*

$$w_r \in \Theta(w_\ell) \tag{5.28}$$

with

$$s(w_\ell, w_r) = s(w_\ell, w_m) = s(w_m, w_r). \tag{5.29}$$

PROOF. Given (5.27) holding with equality, it follows that the Rankine-Hugoniot condition holds for w_ℓ, w_r, with $s(w_\ell, w_r)$ given in (5.29). So $w_r \in \Gamma(w_\ell)$. From (5.19), (5.14) and figure 5.1, (5.26) implies

$$\lambda_+(w_m) \le s(w_m, w_\ell) < \lambda_-(w_\ell) \tag{5.30}$$

whereas from (5.18) and (5.13)

$$\lambda_+(w_r) < s(w_m, w_r) \le \lambda_-(w_m). \tag{5.31}$$

Now (5.28) follows from (5.23), (5.30) and (5.31). □

LEMMA 5.2. *For a given $w_0 \in H$, assume ν_+ strictly monotone on $S_1(w_0)$ and denote by*

$$K(w_0) = \{w \in S_2(w_m) | w_m \in S_1(w_0),\ s(w_0, w_m) < s(w_m, w)\} \tag{5.32}$$

Then $\Omega(w_0) \subseteq K$.

PROOF. From its definition (5.32), the set K is simply connected, so it suffices to determine the boundary. For any w_m the direction of $S_2(w_m)$ is initially the negative ν_--direction, so by the assumption of S_1 strictly monotone in ν_+, for any $w_m \in S_1(w_0) - \{P_2\,(w_0)\}$ there are points in K arbitrarily close to w_m. In particular $S_1(w_0)$ are limits of points in K. Choosing $w_m \in S_1(w_0)$ a sequence approaching w_0, by continuity it is clear that $S_2(w_0)$ are also limits of points in K.

By appeal to lemma 5.1, the inequality in (5.32) fails when $w \in \Theta(w_0)$, for any fixed $w_m \in S_1(w_0)$. By continuity, for w_m close to w_0, the corresponding $w \in \Theta(w_0)$ is close to $P_1(w_0)$, while for w_m close to $P_2(w_0)$, w will be close to $P_2(w_0)$. So again by continuity, all of the overcompressive segment $\Theta(w_0)$ are limits of points in K, and so $\Omega(w_0) \subseteq K(w_0)$. □

Denote by

$$\chi_0(w_0) = \{w = P_0(w_m) | w_m \in R_1(w_0)\} \cup [P_1(w_0), P_2(w_0)] \cup \{w \in R_2(P_2(w_0))\}. \tag{5.33}$$

As shown in figure 5.3, $\chi_0(w_0)$ divides $\mathbb{R}^2$ into two disjoint regions, denoted by $\chi_\pm(w_0)$, with $w_0 \in \chi_+(w_0)$. By inspection of figure 5.3 and appeal to lemma 5.2, we obtain a result on the solvability of the Riemann problem (5.1).

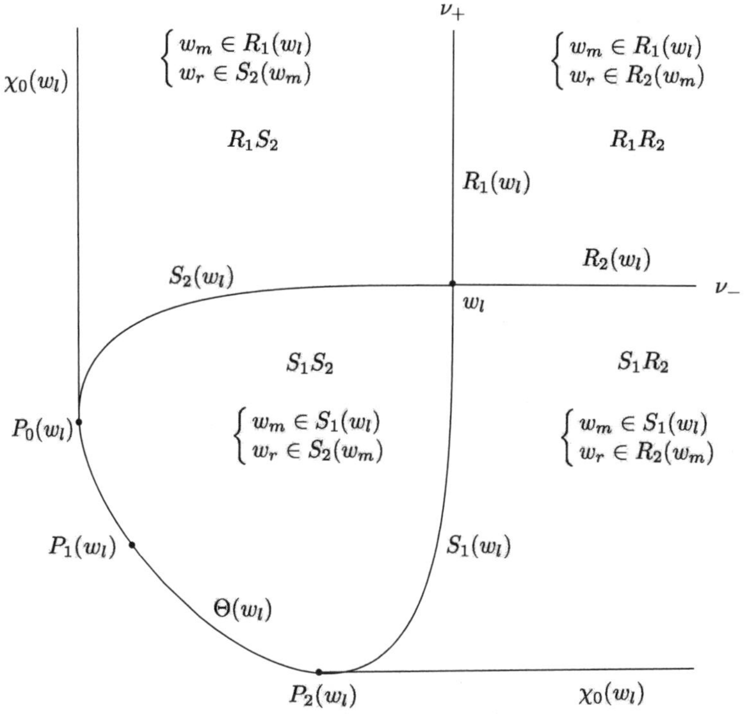

Figure 3.5.3: Classical solution of the Riemann problem

THEOREM 5.3. *Assume ν_+ strictly monotone on $S_1(w_\ell)$. Then the Riemann problem (5.1) is solvable in the class of entropy shocks, overcompressive shocks, and rarefaction waves if*

$$w_r \in \chi_+(w_\ell) \cup \chi_0(w_\ell). \tag{5.34}$$

REMARKS. An entirely similar result, with the roles of the 1- and 2- characteristic families reversed, is obtained by reversing the roles of w_ℓ, w_r in the above discussion.

Uniqueness can fail for a variety of reasons; obviously uniqueness fails if $P_0(w_\ell) \neq P_1(w_\ell)$. In addition, $K(w_\ell)$ may contain points not within $\Omega(w_\ell)$, or uniqueness could fail within the class of "shock-shock" solutions, with $w_r \in \Omega(w_\ell)$. Uniqueness of "rarefaction-shock" solutions is not true in this generality, but is easily obtained by assuming ν_- strictly monotone on $S_2'(w_r)$. We omit the details.

The only possible solutions of (5.1) in this class with $w_r \in \chi_-(w_\ell)$ would have to correspond to $w_r \in K(w_\ell)$, i.e. "shock-shock" solutions. Since $K(w_\ell)$ is bounded, it is clear that the interaction of two or more entropy shocks will result in an unsolvable Riemann problem within this class. Thus for such systems, a wave set $\mathcal{A}_\delta$ containing the entropy shocks, overcompressive shocks and weak entropy-violating shocks (or approximations thereof) cannot be complete.

Such classically unsolvable Riemann problems are solvable, however, in the class of rarefaction waves and singular shocks satisfying (3.41 – 3.43). To this end, we proceed to identify the admissible singular shocks, i.e. the points w_+ satisfying (3.41 – 3.43) for a given $w_- \in H$.

Given $w_- \in H$, we extend the function $s(\cdot, w_-)$ to all of $\Lambda_-(w_-) \cup \Lambda_+(w_-)$ by (3.41), and determine the second component of the Rankine-Hugoniot deficit from (3.6)

$$\begin{aligned} e_2(w_+, w_-) &= s(w_+, w_-)(v_+ - v_-) - g(u_+, v_+) + g(u_-, v_-) \\ &= [(f(u_+, v_+) - f(u_-, v_-))(v_+ - v_-) - (g(u_+, v_+) \\ &\quad - g(u_-, v_-))(u_+ - u_-)] \;/\; (u_+ - u_-) \end{aligned} \tag{5.35}$$

using (3.41).

LEMMA 5.4. *For any $w_- \in H$*

$$e_2(w_+, w_-) > 0, \quad w_+ \in \Lambda_-(w_-) - (\Omega(w_-) \cup \Gamma_-(w_-)), \tag{5.36}$$

$$e_2(w_+, w_-) < 0, \quad w_+ \in \Omega(w_-). \tag{5.37}$$

PROOF. For fixed w_-, the function $e_2(\cdot, w_-)$ given by (5.35) is continuous in w_+ throughout $\Lambda_-(w_-)$ and vanishes only for $w_+ \in \Gamma(w_-)$. Thus (5.35) follows from (1.99) and (5.8). It is here that the sign of the expression in (1.99) is used.

To obtain (5.37), it suffices to show that $e_2(\cdot, w_-)$ changes sign as $\Gamma_-(w_-)$ is traversed at some point $w_+ \in \Gamma_-(w_-)$. Choose any $w_+ \in \Gamma_-(w_-)$ not w_- itself and not the point $P_2(w_-)$. At the point w_+, the curve $\Gamma_-(w_-)$ continues by solution of

$$(q_w(w_+) - s(w_+, w_-))\dot{w}(w_+) = \dot{s}(w_+)(w_+ - w_-) \tag{5.38}$$

dots denoting differentiation along $\Gamma_-(w_-)$. By the choice of w_+, $s(w_+, w_-)$ is not one of the characteristic speeds at w_+ and $w_+ \neq w_-$, so $\dot{s}(w_+) \neq 0$ and the tangent vector to $\Gamma_-(w_-)$ at w_+ is given by

$$T(w_+; w_-) = (q_w(w_+) - s(w_+, w_-))^{-1}(w_+ - w_-)\,. \tag{5.39}$$

For $T_\perp$ orthogonal to $T(w_+; w_-)$, since $u_+ < u_-$, there exists $\hat{s}$ such that

$$(f_u(u_+, v_+) - s(w_+, w_-) \quad f_u(u_+, v_+)\,)T_\perp = \hat{s}(u_+ - u_-), \tag{5.40}$$

which is just the first component of (5.38) with $T_\perp, \hat{s}$ replacing $\dot{w}(w_+)$, $\dot{s}(w_+)$ respectively.

Since the matrix $q_w(w_+) - s(w_+, w_-)$ is nonsingular, this replacement cannot satisfy the second component of (5.38), i.e.

$$(g_u(u_+, v_+) \quad g_v(u_+, v_+) - s(w_+, w_-))T_\perp - \hat{s}(v_+ - v_-) \neq 0. \tag{5.41}$$

Now from (5.40, 5.41), using (5.35), the left side of (5.41) is just $-\nabla e_2(w_+, w_-) \cdot T_\perp$, the nonvanishing of which establishes (5.37). ☐

Thus the condition (3.42) is satisfied by w_+ satisfying the conditions given in (5.36). Next we determine the points w_+ satisfying the weak overcompressibility condition (3.43).

For arbitrary $w_- \in H$, denote by $\mathcal{S}(w_-)$ the set of points $w_+ \in \Lambda_-(w_-)$ satisfying (3.43). The boundary of $\mathcal{S}(w_-)$ is determined by considering (3.43) holding with equality. Thus we consider the two sets

$$\Phi(w_-) \stackrel{\text{def}}{=} \{w_+ \in \Lambda_-(w_-) | s(w_+, w_-) = \lambda_-(w_-)\} \tag{5.42}$$

and

$$\Psi(w_-) \stackrel{\text{def}}{=} \{w_+ \in \Lambda_-(w_-) | s(w_+, w_-) = \lambda_+(w_+)\} \tag{5.43}$$

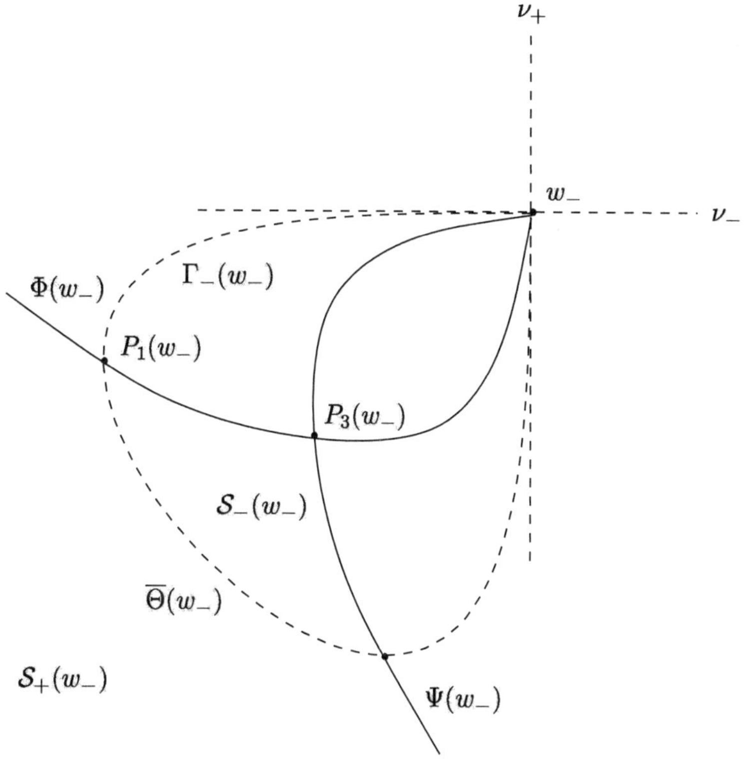

Figure 3.5.4: The trajectories Φ, Ψ

Neither set is empty; from (5.13) and (5.42)

$$\Phi(w_-) \cap \Gamma(w_-) = \{P_1(w_-)\} \tag{5.44}$$

and from (5.14), (5.43)

$$\Psi(w_-) \cap \Gamma(w_-) = \{P_2(w_-)\} \tag{5.45}$$

LEMMA 5.5. *$\Phi(w_-)$ is a single trajectory from w_- to infinity, along which u is monotone, and which approaches w_- in the ν_+-direction.*

PROOF. From (3.41) and (5.42), the points $w_+ \in \Phi(w_-)$ satisfy

$$\lambda_-(w_-)(u_+ - u_-) - f(u_+, v_+) + f(u_-, v_-) = 0 \tag{5.46}$$

We use dots here to denote differentiation along $\Phi(w_-)$, with w_- fixed, thus obtaining from (5.46)

$$(\lambda_-(w_-) - f_u(u_+, u_-))\dot{u}_+ - f_v(u_+, v_+)\dot{v}_+ = 0. \tag{5.47}$$

From (5.7), given an "initial condition", i.e. a point within $\Phi(w_-)$, (5.47) determines v_+ uniquely as a function of $u_+ \in (-\infty, u_-)$. As u_+ approaches u_- with $w_+ \in \Phi(w_-)$, from (5.7) and (5.46) it follows that v_+ must approach v_-. So all such trajectories determined by (5.47) with $w_+ \in \Phi(w_-)$, i.e. satisfying (5.46), contain w_- as a limit point.

But the trajectory determined by (5.47) with w_- as a limit point is unique. Thus $\Phi(w_-)$ is a single trajectory as claimed.

It remains to show how $\Phi(w_-)$ approaches w_-. We rewrite (5.11) in the form

$$\frac{\partial f(u,v)}{\partial \nu_\mp} = \lambda_\pm \frac{\partial u}{\partial \nu_\mp} \tag{5.48}$$

and use (5.48) in (5.47) to obtain

$$(\lambda_-(w_-) - \lambda_-(w_+))\dot{\nu}_+ + (\lambda_-(w_-) - \lambda_+(w_+))\dot{\nu}_- = 0. \tag{5.49}$$

As w_+ approaches w_-, the coefficient of $\dot{\nu}_+$ in (5.49) approaches zero, while that of $\dot{\nu}_-$ does not, using strict hyperbolicity. Thus $|\dot{\nu}_-/\dot{\nu}_+|$ approaches zero in this limit. □

LEMMA 5.6. *$\Psi(w_-)$ is a single trajectory from w_- to infinity, along which both ν_+ and $s(\cdot, w_-) = \lambda_+$ are strictly monotone and unbounded from below. Near w_-, $\Psi(w_-)$ is inside $\Omega(w_-)$ and approaches w_- along the ν_--axis.*

PROOF. Using (3.41) and (5.43), the points $w_+ \in \Psi(w_-)$ satisfy

$$\lambda_+(w_+)(u_+ - u_-) - f(u_+, v_+) + f(u_-, v_-) = 0. \tag{5.50}$$

Using dots here to denote differentiation along $\Psi(w_-)$ with w_- fixed, we obtain from (5.50), using (5.48)

$$\begin{aligned}(\lambda_+(w_+) - \lambda_-(w_+))\frac{\partial u}{\partial \nu_+}(w_+)\dot{\nu}_+ &= (u_- - u_+)\dot{\lambda_+} \\ &= (u_- - u_+)(\frac{\partial \lambda_+}{\partial \nu_+}\dot{\nu}_+ + \frac{\partial \lambda_+}{\partial \nu_-}\dot{\nu}_-).\end{aligned} \tag{5.51}$$

Using (5.12) and strict hyperbolicity in (5.51), it follows that if either $\dot{\nu}_+$ or $\dot{\lambda_+}$ vanishes at some point on a trajectory determined by (5.51), then so does the other. Then using (5.6) in (5.51), $\dot{\nu}_-$ vanishes at such a point, so this is impossible. Thus given any point in $\Psi(w_-)$, such as $P_2(w_-)$, the trajectory determined by (5.51) through the given point determines points in $\Psi(w_-)$ and continues indefinitely or until $\Lambda_0(w_-)$ is approached.

Using (5.7) and (5.50), $\Psi(w_-)$ can approach $\Lambda_0(w_-)$ only at the point w_-. Next we show that all solutions of (5.50, 5.51) in fact do so.

For any fixed

$$\tilde{\lambda} > \lambda_+(w_-) \tag{5.52}$$

denote by

$$\tilde{\Phi} = \{w_+ \in \Lambda_-(w_-) | s(w_+, w_-) = \tilde{\lambda}\}. \tag{5.53}$$

Analogously with (5.49), $\tilde{\Phi}$ is the trajectory determined by

$$(\tilde{\lambda} - \lambda_+(w_+))\dot{\nu}_- + (\tilde{\lambda} - \lambda_-(w_+))\dot{\nu}_+ = 0 \tag{5.54}$$

dots here denoting differentiation along $\tilde{\Phi}$, away from w_-, with w_- as a limit point. In a neighborhood of w_-, from (5.52), (5.54)

$$0 < -\dot{\nu}_+ / \dot{\nu}_- < 1, \quad \dot{\nu}_- < 0 < \dot{\nu}_+ \tag{5.55}$$

and we claim that (5.55) is true indefinitely.

In particular, suppose that $\dot{\nu}_+ \downarrow 0$ at some point. Then from (5.54), $\tilde{\lambda} - \lambda_+(w_+) \downarrow 0$ so $\lambda_+(w_+)$ is nondecreasing in a neighborhood of such a point. But in such a neighborhood

$$\dot{\lambda_+} = \frac{\partial \lambda_+}{\partial \nu_+} \dot{\nu}_+ + \frac{\partial \lambda_+}{\partial \nu_-} \dot{\nu}_- \tag{5.56}$$
$$< 0$$

from (5.6) and (5.55), so this is impossible. Thus there are no points w_+ satisfying (5.50) on $\tilde{\Phi}$ for any $\tilde{\lambda}$ satisfying (5.52). Thus for any $w_+ \in \Psi(w_-)$

$$\lambda_+(w_+) \leq \lambda_+(w_-) \tag{5.57}$$

so all solutions of (5.50, 5.51) indeed approach w_- as a limit point.

Using strict hyperbolicity, for w_+ in a neighborhood of w_-, (5.51) determines $\dot{\nu}_+$ as a smooth function of $w_+, \dot{\nu}_-$, so the trajectory determined by (5.51) with w_- as a limit point is unique in such a neighborhood, proving that $\Psi(w_-)$ is a single trajectory as claimed.

From (5.51), it follows immediately that $\dot{\nu}_+/\dot{\nu}_- \to 0$ as w_- is approached. The claim that $\Psi(w_-)$ is inside $\Omega(w_-)$ near w_- is now apparent from (5.45). □

LEMMA 5.7. *There exists a unique point $P_3(w_-)$ satisfying*

$$P_3(w_-) \in \Omega(w_-) \tag{5.58}$$

and

$$\Phi(w_-) \cap \Psi(w_-) = \{P_3(w_-)\} \tag{5.59}$$

PROOF. For $w_+ \in \Psi(w_-)$ close to w_-, we have $s(w_+, w_-) = \lambda_+(w_+) \approx \lambda_+(w_-)$ and $w_+ \in \Omega(w_-)$ by appeal to lemma 5.5. For $w_+ = P_2(w_-) \in \Psi(w_-)$, from (5.43)

$$\begin{aligned} s(w_+, w_-) &= s(P_2(w_-), w_-) \\ &= \lambda_+(P_2(w_-)) \\ &< \lambda_-(w_-) \end{aligned}$$

by inspection of figure 5.1. From (5.43) and (5.45), by continuity there exists a point $P_3(w_-) \in \Psi(w_-)$ satisfying (5.58) and

$$s(P_3(w_-), w_-) = \lambda_-(w_-) \tag{5.60}$$

implying (5.59). Uniqueness follows from the strict monotonicity of $s(\cdot, w_-)$ on $\Psi(w_-)$, established in lemma 5.6. □

LEMMA 5.8. *The region $\mathcal{S}(w_-)$, containing the points $w_+ \in \Lambda_-(w_-)$ satisfying (3.43), is the closed unbounded region bounded by $\Phi(w_-)$ and $\Psi(w_-)$.*

PROOF. The trajectories $\Phi(w_-)$, $\Psi(w_-)$ continue indefinitely in $\Lambda_-(w_-)$, have w_- as a common limit point, and intersect at one point $P_3(w_-)$ by appeal to lemma 5.7. From (5.42), (5.43), $\Phi(w_-)$, $\Psi(w_-)$ contain the points which satisfy (3.43) with equality, and therefore determine the boundary of $\mathcal{S}(w_-)$. The region $\Lambda_-(w_-)$ is divided into four regions by $\Phi(w_-)$, $\Psi(w_-)$, corresponding to each of the two inequalities in (3.43) satisfied or failing. Since the closed overcompressive segment $\bar{\Theta}(w_-) \subset \mathcal{S}(w_-)$ and $P_3(w_-) \in \Omega(w_-)$ by lemma 5.7, $\mathcal{S}(w_-)$ is the unbounded region as shown in figure 5.4. □

We partition $\mathcal{S}(w_-)$ into three disjoint regions

$$\mathcal{S}(w_-) = \mathcal{S}_+(w_-) \cup \bar{\Theta}(w_-) \cup \mathcal{S}_-(w_-) \tag{5.61}$$

with $\mathcal{S}_+(w_-)$ exterior to $\Gamma_-(w_-)$ and $\mathcal{S}_-(w_-) \subset \Omega(w_-)$. From (5.61), lemma 5.4 and lemma 5.8, w_+, w_- satisfy (3.41–3.43) if and only if $w_+ \in \mathcal{S}_+(w_-)$. Inclusion of these singular shocks suffices to solve the Riemann problem (5.1), but may result in uniqueness failure. In comparison with (5.33), we denote by

$$\begin{aligned}\chi_0'(w_0) = \{w = &P_1(w_m) | w_m \in R_1(w_0)\} \cup \bar{\Theta}(w_0) \\ &\cup \{w \in R_2(P_2(w_0))\}\end{aligned} \tag{5.62}$$

which coincides with $\chi_0(w_0)$ if

$$P_0(w_0) = P_1(w_0). \tag{5.63}$$

We partition $\mathbb{R}^2$ into three disjoint regions

$$\mathbb{R}^2 = \chi_-'(w_0) \cup \chi_0'(w_0) \cup \chi_+'(w_0) \tag{5.64}$$

with

$$\chi_-(w_0) \subseteq \chi_-'(w_0) \tag{5.65}$$

$$\chi_+'(w_0) \subseteq \chi_+(w_0) \tag{5.66}$$

holding with equality if and only if (5.63) holds.

THEOREM 5.9. *The Riemann problem (5.1) with $w_r \in \chi_-'(w_\ell)$ is uniquely solvable in the class of rarefaction waves and singular shocks satisfying (3.41 – 3.43).*

REMARKS. If (5.63) is not true, from (5.65) there are Riemann problems with both classical solutions and "singular shock solutions". However, in this case the classical solution was already not unique.

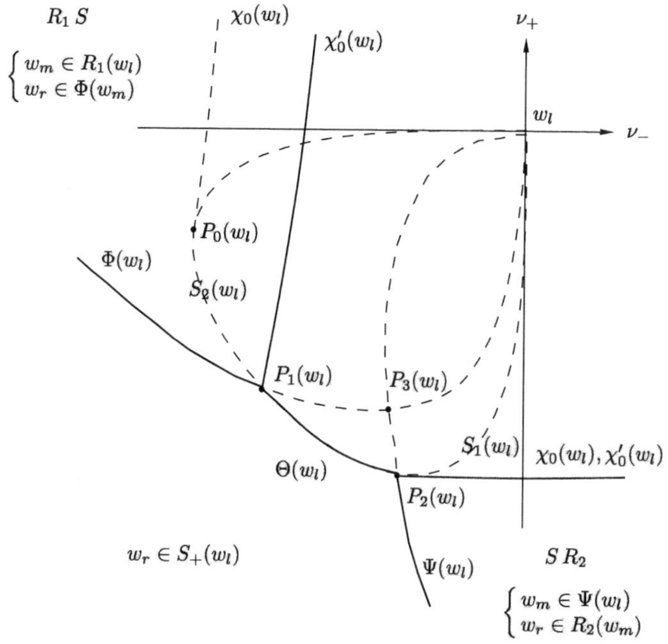

Figure 3.5.5: Singular shock solution of Riemann problem or extended Riemann problem

PROOF. The solution is shown in figure 5.5, which is almost self-explanatory. Indeed, it remains only to prove uniqueness in the "rarefaction-singular shock" case, in which

$$w_m \in R_1(w_\ell), \tag{5.67}$$

$$w_r \in \Phi(w_m). \tag{5.68}$$

Analogously with (5.43), denote by

$$\Psi'(w_-) = \{w_+ \in \Lambda_+(w_-) | s(w_+, w_-) = \lambda_-(w_+)\}. \tag{5.69}$$

An argument entirely similar to the proof of lemma 5.6 shows that ν_- is strictly monotone and unbounded from above on $\Psi'(w_-)$. Since from (5.24) and (5.67),

$$\nu_-(w_m) = \nu_-(w_\ell) \tag{5.70}$$

it follows that w_m is indeed uniquely determined. □

Singular shocks satisfying (3.41–3.42) will accumulate a singular mass in the set

$$\mathcal{M} = \left\{ \begin{pmatrix} 0 \\ \underline{\mathrm{M}} \end{pmatrix}, \ \underline{\mathrm{M}} > 0 \right\} \tag{5.71}$$

the same set as obtained in theorem 1.1 above, under different but compatible structured assumptions, and obtained ad hoc in section 2 for the systems (2.16–2.17) and (2.18–2.19). Completeness of a wave set $\mathcal{A}_\delta$ including such singular shocks thus requires solution of extended Riemann problems (5.2) with this set $\mathcal{M}$ and w_ℓ, w_r as arising from the possible wave interactions.

For $w_r \in \chi'_-(w_\ell)$, the solution of (5.1) contains a single singular shock, with singular mass initially zero. In this case the same solution satisfies (5.2), with M_0 the initial singular mass. For systems such as the model system (2.17–2.18), it is true that the interaction of a singular shock with $w_+ \in \mathcal{S}_+(w_-)$ with a shock (entropy, overcompressive, or singular) results in an initial value problem (5.2) with $w_r \in \mathcal{S}_+(w_\ell)$, i.e. with solution a single singular shock. However, even for the model system the interaction of such a singular shock with a weak entropy-violating shock can result in an initial value problem (5.2) with $M_0 \neq 0$ and

$$w_r \in \chi'_0(w_\ell), \tag{5.72}$$

or

$$w_r \in \chi'_+(w_\ell). \tag{5.73}$$

In [KLS], this problem is solved by an expansion of the set of admissible singular shocks, dropping the condition (3.42) in particular. This can be done more generally.

For example, the proofs of theorems 3.1 and 3.3 hold (indeed the proof of lemma 3.6 is somewhat simplified) if the Rankine-Hugoniot deficit $e = 0$ in (3.6).

Such discontinuities, of the form (3.36) with $M_0 \in \mathcal{M}$ nonzero, $e = 0$ ($\bar{t}$ is immaterial) and satisfying (3.43) correspond to (weakly) overcompressive shocks with nonzero singular mass. So these discontinuities have the same viscous structure, obtained from theorems 3.1 and 3.3, as do singular shocks satisfying (3.41–3.43). Indeed, perhaps better; under the assumptions of theorem 1.6, there is a viscous structure for overcompressive shocks with nonzero singular mass satisfying (1.12) and not just (1.3). This solves the case (5.72).

THEOREM 5.10. *The extended Riemann problem (5.2), (5.72), with nonzero $M_0 \in \mathcal{M}$ given in (5.71) is uniquely solvable in the class of weakly overcompressive shocks with nonzero singular mass and rarefaction waves.*

PROOF. The existence of such a solution is clear from (5.62) or by inspection of figure 5.5. The uniqueness proof is the same as that of theorem 5.9. □

The more general case (5.73) is addressed in the context of a paradox: consider the Cauchy problem for a system admitting singular shocks and equipped with a strictly convex, nonnegative entropy density U vanishing at $w = 0$. Given smooth initial data of compact support but sufficiently large oscillation, singular shocks presumably form and accumulate singular mass. Let us assume that such solutions indeed arise as weak limits as $\varepsilon \downarrow 0$ of solutions w^ε of the regularized problem (1.12), with identity viscosity matrix for simplicity. Then the approximations w^ε satisfy

$$\frac{d}{dt}\int_{-\infty}^{\infty} U(w^\varepsilon(x,t))dx = -\varepsilon \int_{-\infty}^{\infty} w_x^\varepsilon(x,t)\cdot U_{ww}(w^\varepsilon(x,t))w_x^\varepsilon(x,t)dx \tag{5.74}$$

with $\int_{-\infty}^{\infty} U(w^\varepsilon(x,t))dx$ thus nonincreasing as a function of time and bounded above by the initial data.

A simple scaling argument shows that the right side of (5.74) is bounded away from zero for any value of t such that approximations of singular shocks (indeed any shocks if strength bounded away from zero) are present in $w^\varepsilon(\cdot,t)$. So the singular shocks must disappear after some finite time. How? What happens to the accumulated singular mass, which cannot simply disappear?

This question is resolved in [KLS] for the model system by the inclusion in $\mathcal{A}_\delta$ of a class of "evanescent shocks", solutions of the form (3.36) with nonzero $M_0 \in \mathcal{M}$ given by (5.71), $w_+ \in \mathcal{S}_-(w_-)$, and

$$\bar{t} = \underline{\mathrm{M}} \quad / \mid \quad (s(w_+,w_-)(v_+ - v_-) - g(u_+,v_+) + g(u_-,v_-)|. \tag{5.75}$$

Evanescent shocks thus satisfy (3.41) and (3.43), but not (3.42). They nonetheless admit viscous structure.

THEOREM 5.11. *A measure w^0 of the form (3.36), with nonzero $M_0 \in \mathcal{M}$ given in (5.71), $w_+ \in \mathcal{S}_-(w_m)$ and $t < \bar{t}$ given in (5.75), is the weak limit (in the space of measures on $\mathbb{R}$, pointwise in t, as $\varepsilon \downarrow 0$) of a sequence w^ε satisfying (1.1 – 1.3), with identity viscosity matrix A. The corresponding z^ε obtained from (3.8) satisfies (3.9).*

REMARK. With $\bar{t}$ given by (5.75), w^0 given by (3.36) is self-similar, depending only on $(x - st)\;/\;(\bar{t} - t)$. In this sense evanescent shocks correspond to singular shocks with time reversed.

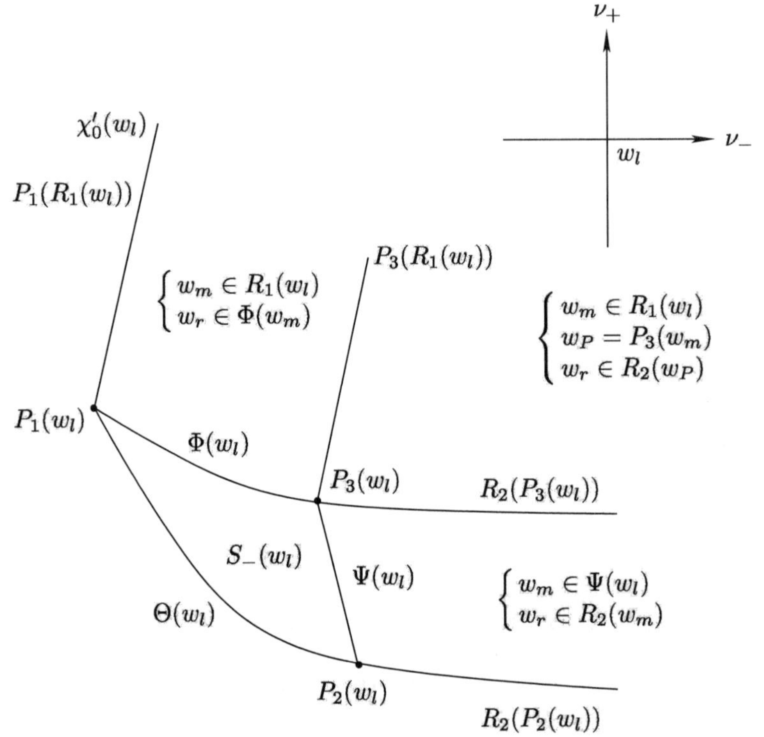

Figure 3.5.6: Solution of extended Riemann problem for $w_r \in \chi_+'(w_\ell)$

PROOF. The argument of theorem 3.3 depends on a positive second component of the singular mass M, not the Rankine-Hugoniot deficit $e = dM/dt$. Indeed, if the inequality (3.42) is reversed, it suffices to replace e_2 by $-e_2$ in (3.64). The proof of lemma 3.6 is unaffected. However, the conclusion of the theorem is now

$$\int_{\mathbb{R}} y_\tau(\xi, \tau)d\xi = -e + o(1) \quad \text{as } \tau \to \infty \tag{5.76}$$

replacing (3.16).

Given (5.76) instead of (3.16), the argument of theorem 3.1 is modified to arrive at the same conclusions. In particular, (3.18) is replaced by

$$w^\varepsilon(x,t) = y\left(\frac{x - st}{\varepsilon}, \frac{\bar{t} - t}{\varepsilon}\right) \tag{5.77}$$

and

$$\tau = \frac{\bar{t} - t}{\varepsilon} \tag{5.78}$$

in (3.19). Replacing y_τ by $-y_\tau$ throughout (3.20 – 3.24), one uses (5.76) to see that the last line of (3.22) is unchanged, and the first right-hand term of (3.24) remains $o(1)$ as $\varepsilon \downarrow 0$. Thus the conclusions of theorem 3.1 remain valid. □

The viscous structure of an evanescent shock is lost as $t \uparrow \bar{t}$, implying singular mass reduced to zero. At $t = \bar{t}$, $w^0(\cdot, t)$ becomes a simple jump discontinuity so continuation in time is simply by solution of a Riemann problem (5.1) with

$w_+ \in \mathcal{S}_-(w_-)$. As $\mathcal{S}_-(w_-) \subset \Omega(w_-)$, by appeal to lemma 5.2 there exists a classical solution containing two entropy shocks.

Thus the fate of singular shocks. Unless the asymptotic conditions at $x = \pm\infty$ support their existence indefinitely, interaction with rarefaction waves will drive a singular shock into the graveyard $\mathcal{S}_-$. They may escape, at least temporarily, by interaction with other shocks. Ultimately, upon exhaustion of its singular mass, an evanescent shock fissions spontaneously into two entropy shocks.

Lest this sound unduly morbid, our story has a happy ending.

THEOREM 5.12. *The extended Riemann problem (5.2) with nonzero $M_0 \in \mathcal{M}$ given in (5.71) and $w_+ \in \chi'_+(w_-)$ is uniquely solvable in the class of evanescent shocks satisfying $w_+ \in \mathcal{S}_-(w_-)$ and rarefaction waves.*

REMARK. This implies that for this class of systems, a wave set $\mathcal{A}_\delta$ containing entropy shocks of any strength, entropy-violating shocks (or suitable approximations thereof) of strength not exceeding δ, overcompressive shocks, with and without singular mass, singular shocks (with $w_+ \in \mathcal{S}_+(w_-)$) and evanescent shocks (with $w_+ \in \mathcal{S}_-(w_-)$) is complete. A front-tracking algorithm based on such a wave set will need to follow the fissioning of evanescent shocks.

PROOF. The solution is shown in figure 5.6. The distinctive feature is a class of "rarefaction-shock-rarefaction" solutions, which is possible because from (5.42), (5.43), (5.59), an evanescent shock connecting w_m on the left with $w_p = P_3(w_m)$ on the right satisfies

$$\lambda_-(w_m) = s(w_m, w_p) = \lambda_+(w_p). \tag{5.79}$$

Uniqueness of the "rarefaction-shock" and "shock-rarefaction" solutions is obtained as in the proof of theorem 5.9. Indeed this derives from lemma 5.6, the result that ν_+ is strictly monotone on $\Psi(w_-)$ and ν_- on $\Psi'(w_-)$ given in (5.69). It remains to prove uniqueness in the "rarefaction-shock-rarefaction" case, the uniqueness of w_m, w_p, for given w_ℓ, w_r, satisfying

$$w_m \in R_1(w_\ell), \tag{5.80}$$

$$w_p = P_3(w_m), \tag{5.81}$$

$$w_r \in R_2(w_p). \tag{5.82}$$

Using (5.24), (5.25), the relations (5.80), (5.82) are equivalent to

$$\nu_-(w_m) = \nu_-(w_\ell), \quad \nu_+(w_m) > \nu_+(w_\ell) \tag{5.83}$$

$$\nu_-(w_r) > \nu_-(w_p), \quad \nu_+(w_r) = \nu_+(w_p). \tag{5.84}$$

We determine the point w_0 from

$$\nu_-(w_0) = \nu_-(w_\ell), \quad \nu_+(w_0) = \nu_+(w_r). \tag{5.85}$$

Now from (5.81), $w_p \in \Psi(w_m)$ along which ν_+ is monotone decreasing away from w_m, and $w_m \in \Psi'(w_p)$ given in (5.69), along which ν_- is monotone increasing away from w_p. So

$$\nu_\pm(w_p) < \nu_\pm(w_m); \tag{5.86}$$

combining (5.83–5.86), we obtain

$$\nu_-(w_p) < \nu_-(w_0), \quad \nu_+(w_m) > \nu_+(w_0). \tag{5.87}$$

Using (5.6) and (5.87)

$$\lambda_+(w_p) < \lambda_+(w_0), \quad \lambda_-(w_m) > \lambda_-(w_0) \tag{5.88}$$

so any solution of (5.80–5.82), which is necessarily also a solution of (5.79, 5.83, 5.84), satisfies an a priori estimate

$$\lambda_-(w_0) < s(w_m, w_p) < \lambda_+(w_0). \tag{5.89}$$

For

$$\lambda \in (\lambda_-(w_0),\ \lambda_+(w_0)) \tag{5.90}$$

we determine

$$w_m(\lambda) = \begin{pmatrix} u_m(\lambda) \\ v_m(\lambda) \end{pmatrix}, \quad w_p(\lambda) = \begin{pmatrix} u_p(\lambda) \\ v_p(\lambda) \end{pmatrix} \tag{5.91}$$

implicitly from

$$\nu_-(w_m(\lambda)) = \nu_-(w_0), \quad \lambda_-(w_m(\lambda)) = \lambda;$$

$$\nu_+(w_p(\lambda)) = \nu_+(w_0), \quad \lambda_+(w_p(\lambda)) = \lambda. \tag{5.92}$$

From (5.6) it is clear that solutions of (5.92), if they exist, are unique. Assuming existence of $w_m(\lambda)$, $w_p(\lambda)$, we consider the first component of the Rankine-Hugoniot deficit $e(\lambda, w_p(\lambda), w_m(\lambda))$ given in $(5.7)_1$ for a system of the form (1.7), denoted here by

$$e_1(\lambda) = \lambda(u_p(\lambda) - u_m(\lambda)) - f(u_p(\lambda), v_p(\lambda)) + f(u_m(\lambda), v_m(\lambda)). \tag{5.93}$$

Now from (3.41) and (5.81), a solution of (5.79, 5.83, 5.84) corresponds to a value $\underline{\lambda}$ such that $e_1(\underline{\lambda}) = 0$, identifying $s(w_m, w_p) = \underline{\lambda}$ in (5.79). We may assume that one such solution exists. Then from (5.6), the solution of (5.92) exists in an open interval

$$I \subseteq (\lambda_-(w_0), \quad \lambda_+(w_0))$$

containing $\underline{\lambda}$. Any additional solutions of (5.79, 5.83, 5.84) correspond to additional zeros of e_1 within I, so it suffices to show that e_1 is strictly monotone within I.

Differentiating (5.93)

$$\begin{aligned}\frac{de_1(\lambda)}{d\lambda} &= u_p(\lambda) - u_m(\lambda) + \left(\lambda \frac{\partial u}{\partial \nu_-}(w_p(\lambda)) - \frac{\partial f}{\partial \nu_-}(w_p(\lambda))\right) \frac{\partial \nu_-}{\partial \lambda_+}(w_p(\lambda)) \\ &\quad - \left(\lambda \frac{\partial u}{\partial \nu_+}(w_m(\lambda)) - \frac{\partial f}{\partial \nu_+}(w_m(\lambda))\right) \frac{\partial \nu_+}{\partial \lambda_-}(w_m(\lambda)) \\ &= u_p(\lambda) - u_m(\lambda) \end{aligned} \tag{5.94}$$

using (5.48), (5.92) and (5.6).

For λ satisfying (5.90), it follows from (5.92), using (5.6), that

$$\nu_-(w_p(\lambda)) < \nu_-(w_0), \quad \nu_+(w_m(\lambda)) > \nu_+(w_0) \tag{5.95}$$

and thus from (5.95) and (5.12)

$$u_p(\lambda) < u_0 < u_m(\lambda). \tag{5.96}$$

The strict monotonicity of e_1 within I now follows from (5.94) and (5.96). □

6. An example

Systems of conservation laws for which distribution solutions are anticipated often fail to be hyperbolic, by reason of an eigenvector deficiency. Examples of such systems admitting delta-shock solutions were given in chapter two. Here we discuss such a system admitting singular shock solutions, hopefully addressing questions which will arise more generally in the context of solvability of the Cauchy problem for systems admitting singular shock solutions.

The system (2.24, 2.25), while closely related to the model system (2.17, 2.18), is nowhere hyperbolic; elementary properties of this system are given in section 2. This system is distinguished, however, by its symmetry group.

THEOREM 6.1. *Among pairs of conservation laws, the system (2.24, 2.25) uniquely is linear in one variable, admitting a symmetry* $(6.42)_1$*, a Galilean symmetry* $(6.43)_1$ *and a scaling symmetry (1.92).*

REMARKS. Our assumption is that (1.92) is a genuine symmetry, not simply a "symmetry at infinity". A similar argument shows that up to elementary scaling, the model system (2.17, 2.18) uniquely admits the symmetries $(6.8)_1$, $(6.42)_1$, $(6.43)_1$ among pairs of conservation laws [Kl, S6].

PROOF. Using $(6.42)_1$, such a system is of the form

$$u_t + (f(u) + c_1 v)_x = 0 \tag{6.1}$$

$$v_t + (g(u) + c_2 v)_x = 0 \tag{6.2}$$

for some constants c_1, c_2. Applying $(6.6)_1$ to (6.1, 6.2) for the symmetry $(6.43)_1$, we obtain

$$f(u) - f(u-\beta) = \beta(1-c)u + \text{const} \tag{6.3}$$

and

$$\begin{aligned} g(u) - g(u-\beta) &= \beta v - c_2\beta u + \beta\big(f(u-\beta) + c_1\,(v - \beta u)\big) + \text{const} \\ &= \beta f(u) - \beta^2 u - c_2\beta u + \beta v(1+c_1) + \text{const} \end{aligned} \tag{6.4}$$

for all $\beta \in \mathbb{R}$, using (6.3) to simplify (6.4). Since (6.4) must hold for all $u, v,$ $c_1 = -1$ and from (6.3)

$$f(u) = u^2 + c_3 u \tag{6.5}$$

for some constant c_3. Now (6.4) can be solved for $g(u)$,

$$g(u) = \frac{1}{3}u^3 + \frac{1}{2}(c_3 - c_2)u^2 + c_4 u \tag{6.6}$$

for some constant c_4. But given the leading terms for $f(u), g(u)$ in (6.5), (6.6), the scaling (1.92) can hold only if $c_2 = c_3 = c_4 = 0$, and with $k = 1, \ell = 2$ in (1.94). □

The linearity $(6.42)_1$ and Galilean $(6.43)_1$ symmetries combine to generate a two-parameter symmetry group $(7.9)_1$ for this system, as for the model system (2.17, 2.18). With $Q(w)$ obtained from $(7.9)_1$, from $(7.35)_1$ we find the characteristic speed u and the single independent Riemann invariant ν given in (2.26) as coordinates for phase space, with respect to which the symmetries correspond to translations.

Rewriting the system (2.24, 2.25) in these variables, we obtain

$$u_t + \big(\frac{u^2}{2} + \nu\big)_x = 0, \tag{6.7}$$

$$\big(\frac{u^2}{2} - \nu\big)_t + \big(\frac{u^3}{3}\big)_x = 0. \tag{6.8}$$

First we address the possible form of solutions of (6.7, 6.8). Introducing a comoving space coordinate y, with dots here denoted $\partial/\partial t + u\partial/\partial x$ and

$$V(y,t) = \frac{\partial x(y,t)}{\partial y}, \quad y \in \mathbb{R}, t \geq 0, \tag{6.9}$$

smooth solution of (6.7, 6.8) satisfy

$$\dot{x} = u \tag{6.10}$$

$$\dot{V} = u_y \tag{6.11}$$

$$\dot{u} = -\nu_y/V \tag{6.12}$$

$$\dot{\nu} = 0, \tag{6.13}$$

assuming $V > 0$. However, from (6.11-6.13) it is clear that in general, even for $V(\cdot, 0)$ positive and bounded away from zero, V will decrease to zero at some point(s) in finite time, precluding the continuation of a smooth solution. There are, of course, examples of continuous solutions continuing indefinitely in time, for example, rarefaction waves connecting (u_-, ν) on the left with (u_+, ν), $u_+ > u_-$, on the right.

The system (2.24, 2.25) is obtained as the system (1.102), applying (1.92-1.93) to the model system (2.17, 2.18). Several properties of the discontinuous solutions are thus determined.

THEOREM 6.2. *Assume a system (1.7, 1.8) satisfying the conditions of section 5. Then the limit system (1.102) obtained by applying (1.92-1.93), satisfies the following.*

For any point $w_0 \in \mathbb{R}^2$

$$\Gamma(w_0) = \{w_0\}. \tag{6.14}$$

The conclusions of theorem 3.3 hold for w_+, w_- *satisfying*

$$u_+ < u_- \tag{6.15}$$

and (3.44), with $s = s(w_+, w_-)$ *obtained from*

$$s(u_+ - u_-) = f^L(u_+, v_+) - f^L(u_-, v_-). \tag{6.16}$$

There are no evanescent shocks for these systems. Assume in addition that the system (1.7) satisfies

$$\begin{pmatrix} \alpha^k \sin\phi \\ \alpha^\ell \cos\phi \end{pmatrix} \in H, \quad \phi \in J \subseteq [0, 2\pi) \tag{6.17}$$

for all sufficiently large α, *with* k, ℓ *as in (1.92). Then for the corresponding system (1.102)*

$$\lambda_\pm(w_0) \in \mathbb{R} \tag{6.18}$$

for all

$$w_0 = \begin{pmatrix} \beta^k \sin\phi \\ \beta^\ell \cos\phi \end{pmatrix}, \quad \phi \in J, \quad 0 < \beta. \tag{6.19}$$

REMARKS. The system (1.102) does not have to be hyperbolic in the region (6.19).

For these systems, singular shocks, having accumulated singular mass, persist indefinitely. The paradox of section 5 does not hold, even if there is a convex entropy density, because $|w_+ - w_-|$ may become arbitrarily small in (3.36).

PROOF. Motivated by $(6.48)_1$ and (1.92), we introduce a class of systems

$$\begin{aligned} u_t + f^\epsilon(u,v)_x &= 0 \qquad f^\epsilon(u,v) = \epsilon^{1+k} f(u/\epsilon^k, v/\epsilon^\ell) \\ v_t + g^\epsilon(u,v)_x &= 0 \qquad g^\epsilon(u,v) = \epsilon^{1+\ell} g(u/\epsilon^k, v/\varepsilon^\ell) \end{aligned} \tag{6.20}$$

with $0 < \epsilon \leq 1$ and k, ℓ as in (1.92). The characteristic speeds and Hugoniot loci for systems (6.20) are denoted by $\lambda^\epsilon, \Gamma^\epsilon$.

The systems (6.20) also satisfy the conditions of section 5, and determine a homotopy between the system (1.7), i.e. with $\epsilon = 1$ in (6.20), and the system (1.102) obtained as the limit as $\epsilon \downarrow 0$.

The claim (6.14) is immediate from (1.99), using the homogeneity conditions (1.97, 1.98). In addition, for any fixed w_0, and any $\epsilon > 0$, $\Gamma^\epsilon(w_0)$ is a bounded connected set, depending continuously on ϵ. Thus

$$\Gamma^\epsilon(w_0) \to \{w_0\} \text{as } \epsilon \downarrow 0. \tag{6.21}$$

Now suppose w_+, w_- are given satisfying (6.15) and (6.16).

Determine $s^\epsilon(w_+, w_-)$ from

$$s^\epsilon(w_+, w_-)(u_+ - u_-) = f^\epsilon(u_+, v_+) - f^\epsilon(u_-, v_-). \tag{6.22}$$

From (6.21), for all sufficiently small ϵ, w_+ is in the exterior of $\Gamma^\epsilon(w_-)$, so by appel to lemma 5.4, using (6.15),

$$s^\epsilon(w_+ w_-)(v_+ - v_-) - g^\epsilon(u_+, v_+) + g^\epsilon(u_-, v_-) > 0 \tag{6.23}$$

From (6.16) and (6.22)

$$s^\varepsilon(w_+, w_-) \to s(w_+, w_-) \text{as } \ \varepsilon \downarrow 0, \tag{6.24}$$

so taking the limit as $\epsilon \downarrow 0$ in (6.23) we find that the condition (3.42) holds for the limit system (1.102).

$$s(w_+, w_-)(v_+ - v_-) - g^L(u_+, v_+) + g^L(u_-, v_-) > 0. \tag{6.25}$$

The left side of (6.25) cannot vanish, because $w_+ \neq w_-$ from (6.15), and so $w_+ \notin \Gamma(w_-)$ from (6.14).

The system (1.102) inherits the conditions (1.92-1.94), (1.99-1.101) from the system (1.7), and satisfies (3.39, 3.40) trivially, using (3.37, 3.38). The assumptions of strict hyperbolicity and genuine nonlinearity at $w_\pm$ are not needed, given the strict overcompressibility condition (3.44).

Thus for the system (1.102), by appeal to theorems 3.3 and 3.1, pairs of points $w_\pm$ satisfying (6.15) and (6.16) are connected by a singular shock with viscous structure. However, using (6.14) and appeal to lemma 5.4, evanescent shocks do not exist for these systems.

Finally assume (6.17) holds, and assume w_0 given of the form (6.19). For all $\epsilon > 0$ sufficiently small, the point

$$\begin{pmatrix} u_0/\epsilon^k \\ v_0/\epsilon^\ell \end{pmatrix} = \begin{pmatrix} (\beta/\epsilon)^k \sin\phi \\ (\beta/\epsilon)^\ell \cos\phi \end{pmatrix} \in H \tag{6.26}$$

from (6.17), so from (6.20)

$$\lambda^\epsilon_\pm(w_0) \in \mathbb{R}. \tag{6.27}$$

As the characteristic speeds are continuous in the limit $\epsilon \downarrow 0$, (6.18) follows form (6.27). □

The singular shocks described in theorem 6.2 are strictly overcompressive, from (3.44). As such, they cannot be combined with rarefaction waves in solutions of initial value problems of the form (5.1) or (5.2). For the system (2.24, 2.25), however, an extension to the case of weak overcompressibility (3.43) is known.

THEOREM 6.3. *For the system (2.24, 2.25), the results of theorem 3.3 hold for $w_\pm$ satisfying*

$$u_+ \le s(w_+, w_-) \le u_-, \tag{6.28}$$

with

$$s(w_+, w_-)(u_+ - u_-) = u_+^2 - v_+ - u_-^2 + v_-. \tag{6.29}$$

REMARKS. (6.29) is (6.16) for the system (2.24, 2.25).

For $u_+ = u_-$, $s(w_+, w_-) = u_\pm$ by convention, and (6.29) requires $v_+ = v_-$. In this the left side of (3.42) vanishes, and we assume nonzero $M \in \mathcal{M}$ given in (5.71).

The condition (6.28) is (3.43) for the system (2.24, 2.25). In terms of the Riemann invariant ν given in (2.26), with $s(w_+, w_-)$ obtained from (6.29), the condition (6.28) becomes

$$|\nu_- - \nu_+| \le \frac{1}{2}(u_- - u_+)^2 \tag{6.30}$$

with $u_+ \le u_-$.

PROOF. The proof of theorem 3.3 is modified, for the particular system (2.24, 2.25), circumventing the use of lemmas 3.5 and 3.7. With $s = \lambda_+(w_+) = u_+$ and $q(y)$ obtained from (2.24, 2.25), (3.54) becomes

$$\begin{aligned} y_{1,\xi} &= y_1(y_1 - u_+) - y_2 + v_+ \\ y_{2,\xi} &= (y_1^3 - u_+^3)/3 - u_+(y_2 - v_+), \quad 2 < \xi < \infty \end{aligned} \tag{6.31}$$

with $y(2, \tau)$ satisfying (3.53) and (3.68), and the point p_1 to be determined. Making the substitutions

$$\begin{aligned} \underline{y}_1 &= y_1 - u_+ \\ \underline{y}_2 &= y_2 - v_+ - u_+(y_1 - u_+) \end{aligned} \tag{6.32}$$

the system (6.31) becomes

$$\begin{aligned} \underline{y}_{1,\xi} &= \underline{y}_1 - \underline{y}_2 \\ \underline{y}_{2,\xi} &= \underline{y}_1^3/3. \end{aligned} \tag{6.33}$$

Introducing the polar coordinates (4.1) for (6.33)

$$\underline{y}_1 = -r\sin\phi$$
$$\underline{y}_2 = r^2\cos\phi \tag{6.34}$$

we obtain from (6.33) a system analogous to (4.2, 4.3),

$$r_\xi = -r^2 T_1(\phi) \tag{6.35}$$
$$\phi_\xi = r T_2(\phi) \tag{6.36}$$

with

$$T_1(\phi) = \sin\phi(\sin^2\phi - \cos\phi + (\sin^2\phi\cos\phi)/3)/(1+\cos^2\phi) \tag{6.37}$$
$$T_2(\phi) = 2(\cos^2\phi - \sin^2\phi\cos\phi + (\sin^4\phi)/6)/(1+\cos^2\phi). \tag{6.38}$$

From (6.37, 6.78), for ϕ_0 satisfying

$$\frac{\cos\phi_0}{\sin^2\phi_0} = \frac{1+\sqrt{1/3}}{2} \tag{6.39}$$

we have

$$T_1(\phi_0) > 0 \tag{6.40}$$
$$T_2(\phi_0) = 0 \tag{6.41}$$
$$T_{2,\phi}(\phi_0) < 0. \tag{6.42}$$

The existence of such ϕ_0 is not accidental, observing that the right side of (6.33) is $q(\underline{y})$ for the system (2.24, 2.25) or $q^L(\underline{y})$ for the system (2.17, 2.18). Thus ϕ_0 obtained in (6.39) is the ϕ_0 assumed in (1.101), as applied to the system (2.17, 2.18).

We choose the point

$$p_1 : r = r_0 > 0, \quad \phi = \phi_0, \tag{6.43}$$

with $r_0 > 0$ determined below, ϕ_0 given in (6.39). Then as $\tau \to \infty$, from (3.53),

$$r(2,\tau) \to r_0, \;\; \phi(2,\tau) \to \phi_0, \tag{6.44}$$

and from (3.68),

$$r_\tau(2,\tau), \; \phi_\tau(2,\tau) \to 0. \tag{6.45}$$

Using (6.40, 6.41, 6.42) and (6.44), it follows by inspection of (6.35), (6.36) that for any fixed, sufficiently large value of τ

$$0 < r(\xi,\tau) < r(2,\tau) \tag{6.46}$$

$$|\phi(\xi,\tau) - \phi_0| \le |\phi(2,\tau) - \phi_0|, \quad 2 < \xi < \infty. \tag{6.47}$$

From (6.47), for $\delta > 0$ to be chosen below and τ sufficiently large, depending on δ we obtain from (6.40), (6.44),

$$1 - \delta \le \frac{T_1(\phi(\xi,\tau))}{T_1(\phi_0)} \le 1 + \delta. \tag{6.48}$$

From (6.44), for all sufficiently large τ,

$$1 - \delta \le \frac{r(2,\tau)}{r_0} \le 1 + \delta. \tag{6.49}$$

Using (6.48, 6.49) in (6.35), we readily obtain

$$\frac{1-\delta}{1+(1+\delta)^2 r_0 T_1(\phi_0)(\xi-2)} \le \frac{r(\xi,\tau)}{r_0} \le \frac{1+\delta}{1+(1-\delta)^2 r_0 T_1(\phi_0)(\xi-2)}, \tag{6.50}$$

$$2 < \xi < \infty,$$

Now (6.33), (6.34), and (6.50) prove (3.14) for this system.

To prove (3.69) for this system in view of (6.33), (6.34), and (6.45), it will suffice to prove that for all sufficiently large τ,

$$\int_2^\infty (|r_\tau(\xi,\tau)| + |\phi_\tau(\xi,\tau)|)d\xi \le c(|r_\tau(2,\tau)| + |\phi_\tau(2,\tau)|), \tag{6.51}$$

with c a generic constant here and below.

For all sufficiently large τ, from (6.44) and (6.47),

$$|T_2(\phi(\xi,\tau)) - (\phi(\xi,\tau)-\phi_0)T_{2,\phi}(\phi_0)| \le \delta|(\phi(\xi,\tau)-\phi_0)T_{2,\phi}(\phi_0)|; \tag{6.52}$$

using (6.52) and the lower bound for $r(\xi,\tau)$ in (6.50) in (6.36), we obtain an estimate

$$\begin{aligned} \left|\frac{\phi(\xi,\tau)-\phi_0}{\phi(2,\tau)-\phi_0}\right| &\le \Big(1+(1+\delta)^2 r_0 T_1(\phi_0)(\xi-2)\Big)^{-\gamma} \\ &\le d(\xi,\tau)^{-\gamma}, \end{aligned} \tag{6.53}$$

introducing the abbreviations

$$d(\xi,\tau) = 1 + T_1(\phi_0) r(2,\tau)(\xi-2), \tag{6.54}$$

$$\gamma = -\frac{(1-\delta)^2\, T_{2,\phi}(\phi_0)}{(1+\delta)^2\, T_1(\phi_0)}, \tag{6.55}$$

and use of (6.49).

We also introduce the function

$$b(\xi,\tau) = r(2,\tau)/d(\xi,\tau) \tag{6.56}$$

satisfying

$$b_\xi(\xi,\tau) = -T_1(\phi_0) b(\xi,\tau)^2 \tag{6.57}$$

and

$$b_\tau(\xi,\tau) = r_\tau(2,\tau)/d(\xi,\tau)^2. \tag{6.58}$$

From (6.56) and (6.14),

$$\omega(\xi,\tau) = r(\xi,\tau) - b(\xi,\tau) \tag{6.59}$$

vanishes at $\xi = 2$ for all τ. In particular, we shall use

$$\omega_\tau(2,\tau) = 0. \tag{6.60}$$

From (6.59), using $r_0 > 0$ in (6.56) and (6.49), it is clear that

$$|\omega(\xi,\tau)| \le b(\xi,\tau) \tag{6.61}$$

for $\xi - 2$ sufficiently small and τ sufficiently large. For ξ such that (6.61) holds, differentiating (6.59) and using (6.35), (6.57), we obtain

$$\omega_\xi = -2T_1(\phi) b\omega - T_1(\phi)\omega^2 + \big(T_1(\phi) - T_1(\phi_0)\big) b^2, \tag{6.62}$$

which becomes, using (6.61)

$$|\omega|_\xi \leq -T_1(\phi)b|\omega| + c(\phi - \phi_0)b^2. \tag{6.63}$$

From (6.63), using (6.48), (6.53), (6.56) and (6.49)

$$|\omega(\xi,\tau)| \leq c|\phi(2,\tau) - \phi_0|r_0 d(\xi,\tau)^{-1-\gamma} \tag{6.64}$$

from which it follows that for τ sufficiently large, (6.61) holds for all ξ and in a stronger form

$$|\omega(\xi,\tau)| \leq \delta b(\xi,\tau). \tag{6.65}$$

We differentiate (6.62) with respect to τ, obtaining

$$\begin{aligned} \omega_{\tau\xi} = &-2T_1(\phi)(b+\omega)\omega_\tau - (b^2 + 2b\omega + \omega^2)T_{1,\phi}(\phi)\phi_\tau \\ &+ 2b(T_1(\phi) - T_1(\phi_0))b_\tau, \end{aligned} \tag{6.66}$$

which is estimated using (6.48) and (6.65) in the first right-hand term, (6.65) and (6.37) in the second, and (6.37), (6.53) in the third.

We thus obtain an estimate

$$\begin{aligned} |\omega_\tau|_\xi &\leq -2(1-\delta)^2 bT_1(\phi_0)|\omega_\tau| + cb^2|\phi_\tau| \\ &\quad + c|r_\tau(2,\tau)|\,|\phi(2,\tau) - \phi_0|\,bd^{-2-\gamma} \\ &\leq -2(1-\delta)^2 r(2,\tau)T_1(\phi_0)\frac{|\omega_\tau|}{d} + cr_0^2\frac{|\phi_\tau|}{d^2} \\ &\quad + cr_0|r_\tau(2,\tau)|\,|\phi(2,\tau) - \phi_0|d^{-3-\gamma} \end{aligned} \tag{6.67}$$

using (6.56) and (6.49) in the second step.

Differentiating (6.36) with respect to τ, using (6.59),

$$\phi_{\tau\xi} = (b+\omega)T_{2,\phi}\phi_\tau + T_2(\phi)(b_\tau + \omega_\tau); \tag{6.68}$$

again for all τ sufficiently large, it is no loss of generality to assume

$$|T_{2,\phi}(\phi)| \geq (1-\delta)|T_{2,\phi}(\phi_0)|. \tag{6.69}$$

Now using (6.69), (6.56), (6.49), (6.65), (6.38), (6.41), (6.53), and (6.58) in (6.68),

$$\begin{aligned} |\phi_\tau|_\xi &\leq (1-\delta)^2 T_{2,\phi}(\phi_0)r(2,\tau)\frac{|\phi_\tau|}{d} + c|\phi(2,\tau) - \phi_0|\frac{|\omega_\tau|}{d^\gamma} \\ &\quad + c|\phi(2,\tau) - \phi_0|\,|r_\tau(2,\tau)|d^{-2-\gamma} \\ &= (1+\delta)^2\gamma T_1(\phi_0)r(2,\tau)\frac{|\phi_\tau|}{d} + c|\phi(2,\tau) - \phi_0|\frac{|\omega_\tau|}{d^\gamma} \\ &\quad + c|\phi(2,\tau) - \phi_0|\,|r_\tau(2,\tau)|d^{-2-\gamma} \end{aligned} \tag{6.70}$$

using (6.55) in the last step. Combining (6.67) and (6.70),

$$\begin{aligned} \frac{1}{2}\frac{\partial}{\partial\xi}(\omega_\tau^2 + \phi_\tau^2) &\leq -\frac{r(2,\tau)T_1(\phi_0)}{d}[2(1-\delta|^2\omega_\tau^2 + (1+\delta)^2\gamma\phi_\tau^2] \\ &\quad + c|\omega_\tau\phi_\tau|\Big(\frac{1}{d^2} + \frac{1}{d^\gamma}\Big) + c|r_\tau(2,\tau)|\big(|\omega_\tau|d^{-3-\gamma} + |\phi_\tau|d^{-2-\gamma}\big). \end{aligned} \tag{6.71}$$

With ϕ_0 obtained explicitly from (6.39), using (6.37), (6.38) we find

$$\frac{|T_{2,\phi}(\phi_0)|}{T_1(\phi_0)} \approx 4.56. \tag{6.72}$$

Therefore choosing δ sufficiently small, we may assume $\gamma \geq 4$ and $(1-\delta)^2 \geq 3/4$ in (6.71), obtaining

$$\frac{1}{2}\frac{\partial}{\partial \xi}(\omega_\tau^2 + \phi_\tau^2) \leq -\frac{3}{2}\frac{r(2,\tau)T_1(\phi_0)}{d}(\omega_\tau^2 + \phi_\tau^2) + \frac{c}{d^2}(\omega_\tau^2 + \phi_\tau^2) \\ + c|r_\tau(2,\tau)|^2 d^{-10}. \tag{6.73}$$

The inequality (6.73) is easily integrated, using (6.54) and (6.60), to obtain

$$\big(\omega_\tau(\xi,\tau)^2 + \phi_\tau(\xi,\tau)^2\big) d(\xi,\tau)^3 \exp\big(\frac{c}{d(\xi,\tau)}\big) \leq c\phi_\tau(2,\tau)^2 \\ + cr_\tau(2,\tau)^2 \int_0^\xi d(\eta,\tau)^{-7} d\eta \tag{6.74}$$

for all $\xi > 2$. For any fixed $r_0 > 0$, for all sufficiently large τ,

$$|\omega_\tau(\xi,\tau)|,\ |\phi_\tau(\xi,\tau)| \leq O(\xi^{-3/2})$$

for large ξ, using (6.74) and (6.54), and (6.51) follows easily from (6.74). □

By appeal to theorems 6.3 and 3.1, for the system (2.24, 2.25), the definition of a singular shock connecting, w_- on the left to w_+ on the right is satisfied for $w_\pm$ satisfying (6.28, 6.29), equivalently (6.30). The corresponding measures (3.36) correspond to admissible singular shocks for the system.

THEOREM 6.4. *For the system (2.24, 2.25), the initial value problems (5.1) and (5.2), with $\mathcal{M}$ as in (5.71), are solvable in the class of admissible singular shocks and rarefaction waves. Uniqueness fails only in the case $\nu_\ell = \nu_r$, $u_\ell < u_r$, $M_0 \neq 0$.*

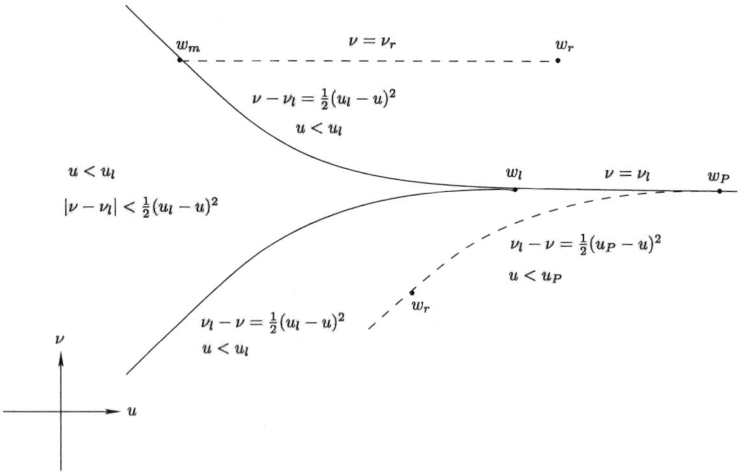

Figure 3.6.1: Solution of Riemann problem and extended Riemann problem

PROOF. The solution is shown in figure 6.1; there are only three generic cases to be considered [SSS].

For

$$|\nu_\ell - \nu_r| \leq \frac{1}{2}(u_\ell - u_r)^2, \tag{6.75}$$

the solution is a single, singular shock connecting w_ℓ to w_r.

For

$$\nu_r - \nu_\ell > \frac{1}{2}(u_\ell - u_r)^2, \tag{6.76}$$

the solution is a singular shock connecting w_ℓ to w_m, satisfying

$$\nu_m = \nu_r \tag{6.77}$$

$$s(w_\ell, w_m) = u_m < u_\ell \tag{6.78}$$

$$\nu_m - \nu_\ell = \frac{1}{2}(u_\ell - u_m)^2 \tag{6.79}$$

and a rarefaction connecting w_m to w_r.

Finally, for

$$\nu_r - \nu_\ell < -\frac{1}{2}(u_\ell - u_r)^2 \tag{6.80}$$

the solution is a rarefaction connecting w_ℓ to w_p, i.e. with

$$\nu_p = \nu_\ell, \; u_p > u_\ell \tag{6.81}$$

and a singular shock connecting w_p to w_r, satisfying

$$s(w_p, w_r) = u_p > u_r \tag{6.82}$$

$$\nu_p - \nu_r = \frac{1}{2}(u_p = u_r)^2. \tag{6.83}$$

For $\nu_r = \nu_\ell$, $u_r > u_\ell$ and $M_0 = 0$, ω_r and ω_ℓ are connected by a rarefaction and there is no singular shock. □

However for $\nu_r = \nu_\ell$, $u_r > u_\ell$ and $M_0 \neq 0$, there may be any number of rarefaction waves connecting w_ℓ to w_r, separated by "degenerate" singular shocks, i.e. with $w_+ = w_-$, $s = u_\pm$, but nonzero singular mass. Thus in this special case there exists a plethora of solutions within this class.

Taking a weak limit in the space of measures as the number of rarefaction waves and singular shocks becomes infinite, we can obtain additional solution candidates of another class, for example

$$w(x,t) = \begin{cases} \begin{pmatrix} u_\ell \\ v_\ell \end{pmatrix}, & x/t \leq u_\ell \\ \begin{pmatrix} x/t \\ \frac{1}{2}(x/t)^2 - \nu_\ell \end{pmatrix} + \begin{pmatrix} 0 \\ \underline{M}/(t(u_r - u_\ell)) \end{pmatrix}, & u_\ell < \frac{x}{t} < u_r \\ \begin{pmatrix} u_r \\ v_r \end{pmatrix}, & \frac{x}{t} \geq u_r \end{cases} \tag{6.84}$$

corresponding to a uniform distribution of the initial Dirac mass.

Next we consider the solution of more general initial value problems for the system (2.24, 2.25) by application of a front-tracking algorithm. Much of the procedure is familiar [BCP]. Modifications are made to allow for initial data of large oscillation, and to allow for the anticipated low regularity of the solution [KLS]. We determine a wave set $\mathcal{A}_\delta$ containing the admissible singular shocks as determined by theorem 3.3, and jump discontinuities

$$w(x,t) = \begin{cases} w_-, & x < st \\ w_+, & x > st \end{cases} \tag{6.85}$$

with $w_\pm, s$ to be determined.

THEOREM 6.5. *For the system (2.24, 2.25), the only useable jump discontinuities (6.85) in $\mathcal{A}_\delta$ correspond to*

$$|u_+ - u_-| \leq \delta \tag{6.86}$$

$$|\nu_+ - \nu_-| \leq o(|u_+ - u_-|) \tag{6.87}$$

$$|s - \frac{1}{2}(u_+ + u_-)| \leq o(|u_+ - u_-|) \tag{6.88}$$

PROOF. Strong jump discontinuities, i.e. not of strength δ, have to satisfy the Rankine-Hugoniot condition (1.9). For the system (2.24, 2.25), this is impossible, from (6.14). Weak jump discontinuities, of strength not exceeding δ, are useable if the Rankine-Hugoniot deficit $e(s, w_+, w_-)$ given in (3.6) satisfies

$$|e(s, w_+, w_-)| \leq o(|w_+ - w_-|). \tag{6.89}$$

For the system (2.24, 2.25), rewritten in the form (6.7, 6.8), using (2.26), we obtain from (3.6)

$$e_1 = s(u_+ - u_-) - (\frac{u_+^2}{2} + \nu_+) + (\frac{u_-^2}{2} + \nu_-) \tag{6.90}$$

$$e_2 = s(\frac{u_+^2}{2} - \nu_+ - \frac{u_-^2}{2} + \nu_-) - \frac{u_+^3}{3} + \frac{u_-^3}{3}. \tag{6.91}$$

Eliminating the speed s from (6.90), (6.91) we obtain

$$(\nu_+ - \nu_-)^2 = (\frac{u_+^2}{2} - \frac{u_-^2}{2})e_1 - (\nu_+ - \nu_-)e_1 - (u_+ - u_-)e_2 - \frac{1}{12}(u_+ - u_-)^4. \tag{6.92}$$

Assuming $|w_+ - w_-| \leq \delta$, (6.86) and (6.87) follow from (6.89) and (6.92). Then (6.88) follows from (6.90), using (6.87) and (6.89). □

For definiteness, we choose for the set $\mathcal{A}_\delta$ the jump discontinuities (6.85) satisfying

$$0 < u_+ - u_- \leq \delta, \qquad s = \frac{1}{2}(u_= + u_-), \quad \nu_+ = \nu_- \tag{6.93}$$

for which, from (6.90, 6.91)

$$e_1 = 0, \ \ e_2 = -(u_+ - u_-)^3/12. \tag{6.94}$$

For the system (2.24, 2.25), rarefaction waves are approximated by finite sequences of jump discontinuities (6.85) satisfying (6.93). Thus by appeal to theorem 6.4, the wave set $\mathcal{A}_\delta$ containing the admissible singular shocks (as determined by theorem 3.3) and the jump discontinuities satisfying (6.93) is complete as defined in section 5.

We may thus attempt to solve the Cauchy problem for the system (2.24, 2.25) by application of a front-tracking algorithm, using this wave set $\mathcal{A}_\delta$. For suitable initial data, proof of convergence of a sequence of front-tracking approximations derives from symmetry group of this system. This system is invariant under translation of u, ν coordinates for phase space. This is reflected in the results of theorem 6.4, with the solution depending only on $u_r - u_\ell$, $\nu_4 - \nu_\ell$. Given such results, control of the variation of front-tracking approximations is readily obtained.

For given initial data and any $\delta > 0$ denote by w^δ the front-tracking approximation obtained using the wave set $\mathcal{A}_\delta$. For any $t \geq 0$, $w^\delta(\cdot, t)$ is piecewise constant in x, with discontinuities corresponding to waves in $\mathcal{A}_\delta$. Assume a finite number $N(\delta, t)$ of points of discontinuity, which we order $x_i < x_{i+1}, i = 1, \ldots, N(\delta, t) - 1$. The wave at x_i moves with speed s_i, connects $w_{i,-} = w_{i-1,+}$ on the left to $w_{i,+} = w_{i+1,-}$ on the right, and has singular mass $M_i \in \mathcal{M}$; by convention, $M_i = 0$ if the wave at x_i is not a singular shock, i.e. of the form (6.85). Of course the $x_i, s_i, w_{i,\pm}, M_i$ depend on δ, t; we ignore this dependence in the notation, in the interest of simplicity.

We measure the total wave strength by

$$var\ u^\delta(\cdot, t) = \sum_{i=1}^{N(\delta,t)} |u_{i,+} - u_{i,-}|, \tag{6.95}$$

$$var\ \nu^\delta(\cdot, t) = \sum_{i=1}^{N(\delta,t)} |\nu_{i,+} - \nu_{i,-}|, \tag{6.96}$$

and

$$M(\delta, t) = \sum_{i=1}^{N(\delta,t} |M_i|, \tag{6.97}$$

the total singular mass. We thus separate the regular and singular parts of v^δ,

$$\begin{aligned} v^\delta &= \tilde{v}^\delta + M_2^\delta \\ \tilde{v}^\delta &= \frac{1}{2}(u^\delta)^2 - \nu^\delta \\ M^\delta(x, t) &= \sum_{i=1}^{N(\delta,t)} M_i \delta(x - x_i), \end{aligned} \tag{6.98}$$

with M_2^δ the second components of M^δ.

LEMMA 6.6. *For the system (2.24, 2.25), the quantities $N(\delta, \cdot)$, $var\ u^\delta, var\ \nu^\delta$ are nonincreasing in t as a result of wave interaction, using the wave set $\mathcal{A}_\delta$, for any fixed $\delta > 0$.*

PROOF. The wave set $\mathcal{A}_\delta$ contains only singular shocks, satisfying (6.28, 6.29) and jump discontinuities satisfying (6.93). This limits the possible wave interactions, and one has only to check the solution obtained from theorem 6.4 in each case.

The interaction of two or more singular shocks, each satisfying (6.28, 6.29, 6.30) results in an initial value problem (5.2) satisfying (6.75). The solution is thus a single singular shock. Such an interaction decreases $N(\delta, \cdot)$, and leaves $var\ u^\delta$ unchanged; $var\ \nu^\delta$ may decrease or be unchanged.

Two or more waves of the form (6.85), (6.93) cannot interact, as their speeds necessarily are increasing with respect to x.

The interaction of a singular shock and a jump discontinuity produces a singular shock and possibly a jump discontinuity. In this case $N(\delta,\cdot)$ and $var\ u^\delta$ are unchanged, or decrease, and $var\ \nu^\delta$ is unchanged.

Finally, a singular shock can interact simultaneously with two jump discontinuities (6.93), one on each side. In this case, the solution is a singular shock and up to two jump discontinuities, on the same side of the singular shock. Again $N(\delta,\cdot)$ and $var\ u^\delta$ are unchanged or decrease and $var\ \nu^\delta$ is unchanged by such interaction.

The case where uniqueness fails in theorem 6.4 is possible, if a singular shock satisfying $u_- - u_+ < 2\delta$ and $\nu_+ = \nu_-$ interacts with two sufficiently strong jump discontinuities in this manner. In such a case we choose a solution containing a single singular shock, and one or two jump discontinuities, as required from (6.93). □

We note that $M(\delta,\cdot)$ given in (6.97) is unaffected by wave interaction.

LEMMA 6.7. *For the system (2.24, 2.25), using the wave set $\mathcal{A}_\delta$, $var\ u^\delta(\cdot,0)$ and $var\ \nu^\delta(\cdot,0)$ will be bounded uniformly with respect to δ if and only if the initial data is of bounded variation, with $\nu(\cdot,0)$ piecewise constant and satisfying*

$$\sum_i |\nu(x_i+0,0)-\nu(x_i-0,0)|^{\frac{1}{2}} < \infty, \tag{6.99}$$

with x_i the points of discontinuity of $\nu(\cdot,0)$.

PROOF. The necessity of initial data of bounded variation is clear.

Each jump of ε in the value of $\nu^\delta(\cdot,0)$ corresponds to a singular shock in $w^\delta(\cdot,0)$, corresponding to a jump in $u^\delta(\cdot,0)$ not less than $(2|\varepsilon|)^{\frac{1}{2}}$ from (6.30). This establishes the necessity and sufficiency of the condition (6.99), assuming that $\nu(\cdot,0)$ is piecewise constant. Thus it suffices to show that $\nu(\cdot,0)$ has to be piecewise constant. Suppose not; then there is a finite interval J within which $\nu(\cdot,0)$ is strictly monotone.

The piecewise constant function $\nu^\delta(\cdot,0)$ satisfies

$$|\nu^\delta(\cdot,0)-\nu(\cdot,0)| \le \varepsilon_\delta \text{ a.e. in } J$$

for some ε_δ such that

$$\varepsilon_\delta \downarrow 0 \text{ as } \delta \downarrow 0. \tag{6.100}$$

For $\nu(\cdot,0)$ strictly monotone in J, there must be a large number of small jumps in $\nu^\delta(\cdot,0)$. In particular, for $\{x_i\}$ the points of discontinuity of $\nu^\delta(\cdot,0)$ and

$$\varepsilon_i = |\nu^\delta(x_i+0,0)-\nu^\delta(x_i-0,0)|$$

there is a constant c, independent of δ and ε_δ, such that

$$\sum_{\substack{x_i\in J\\ \varepsilon_i\le 2\varepsilon_\delta}} \varepsilon_i \ge c > 0 \tag{6.101}$$

From (6.30), as discussed above, at each $x_i \in J$ the variation in $u^\delta(\cdot, 0)$ is of order $O(\varepsilon_i^{\frac{1}{2}})$, so

$$\begin{aligned}\sum_{x_i \in J} |u^\delta(x_i + 0, 0) - u^\delta(x_i - 0, 0)| \\ \geq c \sum_{x_i \in J} \varepsilon_i^{\frac{1}{2}} \\ \geq c/\varepsilon_\delta^{\frac{1}{2}}\end{aligned} \tag{6.102}$$

which increases without bound as $\delta \downarrow 0$ from (6.100). □

LEMMA 6.8. *For each $\delta > 0$, assume μ^δ a piecewise constant function of x, t in $\mathbb{R} \times \mathbb{R}_+$, for each δ, t satisfying*

$$\mu^\delta(x, t) = \mu_i, \quad x_i < x < x_{i+1}, \quad i = 1, \cdots, N(\delta, t) < \infty \tag{6.103}$$

with $x_1 = -\infty$ and $x_{N(\delta,t)+1} = +\infty$ by convention. The x_i, μ_i are renumbered at each t where two or more of the x_i separate or coalesce. Elsewhere, the μ_i are constants, depending on δ, and the x_i are differentiable in t, with

$$s_i = \frac{dx_i}{dt} \tag{6.104}$$

bounded uniformly with respect to δ, t.

Assume

$$var\mu^\delta(\cdot, t) = \sum_{i=1}^{N(\delta,t)-1} |\mu_{i+1} - \mu_i| \tag{6.105}$$

bounded uniformly with respect to δ, t, with μ_1 and $\mu_{N(\delta,t)}$ independent of δ, t.

Then for any $1 \leq p < \infty$ there exists a convergent subsequence (also denoted by $\{\mu^\delta\}$), with limit μ^0 a bounded measurable function on $\mathbb{R} \times \mathbb{R}_+$,

$$\mu^\delta(\cdot, t) \overset{\delta \downarrow 0}{\rightarrow} \mu^0(\cdot, t) \tag{6.106}$$

in $L^p(\mathbb{R})$, uniformly with respect to t in bounded intervals; and with ∂ either $\partial/\partial x$ or $\partial/\partial t$, $1 \leq \tilde{p} \leq p$

$$\partial(\mu^\delta)^{\tilde{p}} \overset{\delta \downarrow 0}{\rightharpoonup} \partial(\mu^0)^{\tilde{p}} \tag{6.107}$$

weakly in the space of measures on $\mathbb{R} \times \mathbb{R}_+$.

PROOF. For fixed $p < \infty$, it follows from (6.105) that there exists a subsequence $\{\mu^\delta\}$ such that (6.106) holds for any rational value of $t \in [0, 1)$. Using the uniform boundedness of the s_i in (6.104), for μ^δ of this form one readily verifies that

$$\frac{d}{dt} \int_{-\infty}^{\infty} |\mu^\delta(x, t) - \mu^{\delta'}(x, t)|^p dx \leq c \tag{6.108}$$

uniformly with respect to t, δ, δ'. Thus (6.106) holds uniformly with respect to $t \in [0, 1)$, and the standard diagonalization argument extends this to $t \in \mathbb{R}_+$.

From (6.108), the μ^δ are uniformly bounded, so the limit μ^0 is a bounded measurable function.

For ∂ either $\partial/\partial x$ or $\partial/\partial t$, using (6.105) and s_i uniformly bounded in (6.104), it follows that $\partial(\mu^\delta(\cdot,t)^{\tilde p})$ are bounded measures on $\mathbb{R}$, for any finite $\tilde p$, uniformly with respect to t,δ. Then for $\tilde p \le p$ and $\theta \in C_0^\infty(\mathbb{R}\times\mathbb{R}_+)$

$$\iint \theta\partial((\mu^\delta)^{\tilde p})dx\,dt = -\iint \partial\theta(\mu^\delta)^{\tilde p}dx\,dt$$

$$\stackrel{\delta\downarrow 0}{\to} -\iint \partial\theta(\mu^0)^{\tilde p}dx\,dt, \tag{6.109}$$

observing in the last step that as $x_1, x_{N(\delta,t)}$ are bounded for any bounded t, convergence of μ^δ in L_p implies convergence in $L_{\tilde p}$. Now the limit of the left side of (6.109) is defined as θ becomes merely continuous and of compact support, thus establishing (6.107). □

With these preliminaries, we obtain the existence and uniqueness of solutions of a restricted class of Cauchy problems for the system (2.24, 2.25) by application of a front-tracking algorithm.

THEOREM 6.9. *For the system (2.24, 2.25), assume initial data of bounded variation, with $\nu(\cdot,0)$ piecewise constant and satisfying (6.99). Then there exists a convergence sequence as $\delta\downarrow 0$ of front tracking approximations w^δ. For each $\delta>0$, w^δ continues indefinitely in t, with $w^\delta(\cdot,t)$ a locally bounded measure on $\mathbb{R}$. Denoting the limit $\underline{w}$ by*

$$\underline{w} = \begin{pmatrix}\underline{u}\\ \underline{v}\end{pmatrix},\quad \underline{v} = \tilde v + \tilde M,\quad \tilde v = \frac{1}{2}\,\underline{u}^2 - \underline{\nu}, \tag{6.110}$$

the functions $\underline{w}, \tilde v$ and $\underline{\nu}$ are bounded and measurable on $\mathbb{R}\times\mathbb{R}_+$; $\underline{u}_t, \underline{u}_x, (\underline{u}^2)_x, (\underline{u}^3)_x, \tilde v_t, \tilde v_k, \underline{\nu}_t, \underline{\nu}_x$ are bounded measures on $\mathbb{R}$, uniformly in t; $\tilde M$ is a nonnegative bounded measure on $\mathbb{R}$, pointwise in t, and satisfies

$$\int_{-\infty}^{\infty} |\tilde M(x,t)|dx \le ct \tag{6.111}$$

Finally, the limit $\underline{w}$ satisfies

$$\underline{u}_t + (\underline{u}^2 - \tilde v)_x = 0 \tag{6.112}$$

weakly in the space of measures on $\mathbb{R}\times\mathbb{R}_+$, and

$$(\tilde v + \tilde M)_t + \frac{1}{3}(\underline{u}^3)_x = 0 \tag{6.113}$$

weakly in the dual of the space of functions θ, continuous and of compact support in $\mathbb{R}\times\mathbb{R}_+$, θ_x continuous and vanishing in an open subset of $\mathbb{R}\times\mathbb{R}_+$ containing the closed support of $\tilde M$.

PROOF. Given initial data of bounded variation, $N(\delta,0)$ is finite for all $\delta>0$. Lemma 6.6 implies that the corresponding front-tracking approximations w^δ continue indefinitely in t, possibly not uniquely.

From the condition (6.99) on the initial data, lemma 6.7, and lemma 6.6, the w^δ satisfy

$$var\ u^\delta(\cdot,t) \le c \tag{6.114}$$

and

$$var\ \nu^\delta(\cdot,t) \le c, \tag{6.115}$$

in both cases uniformly with respect to t and δ.

The accumulation of singular mass in the front-tracking approximations is determined from $(3.59)_1$. For a singular shock at x_i speed s_i, and limiting values $w_{i,\pm} = \binom{u_{i,\pm}}{v_{i,\pm}}$, the corresponding singular mass M_i satisfies, from (2.24, 2.25),

$$\frac{dM_i}{dt} = \begin{pmatrix} s_i(u_{i,+} - u_{i,-}) - u_{i,+}^2 + u_{i,-}^2 + v_{i,+} - v_{i,-} \\ s_i(v_{i,+} - v_{i,-}) - \frac{1}{3}u_{i,+}^3 + \frac{1}{3}u_{i,-}^3 \end{pmatrix} \tag{6.116}$$

with the first component vanishing from (3.41), or (6.29) and the second nonnegative from (3.42). With $u_{i,-} > u_{i,+}$ and s_i so determined, using

$$\nu_{i,\pm} = \frac{1}{2}u_{i,\pm}^2 - v_{i,\pm}, \tag{6.117}$$

the second component of (6.116) is estimated using (6.30),

$$\begin{aligned} s_i(\frac{1}{2}u_{i,+}^2 - \frac{1}{2}u_{i,-}^2 - \nu_{i,+} + \nu_{i,-}) - \frac{1}{3}u_{i,+}^3 + \frac{1}{3}u_{i,-}^3 \\ = \frac{(\nu_{i,+} - \nu_{i,-})^2}{u_{i,-} - u_{i,+}} + \frac{(u_{i,-} - u_{i,+})^3}{12} \\ \leq \frac{1}{2}(u_{i,-} - u_{i,+}) + \frac{1}{12}(u_{i,-} - u_{i,+})^3 \\ \leq c(u_{i,-} - u_{i,+}), \end{aligned} \tag{6.118}$$

since the $u_{i,\pm}$ are point values of the uniformly bounded function u^δ from (6.114). Thus the total singular mass $M(\delta, t)$ satisfies, uniformly with respect to δ,

$$M(\delta, t) \leq ct \tag{6.119}$$

from (6.97), (6.116), (6.95), (6.118) and (6.114). In particular M^δ obtained from (6.98) is a bounded measure on $\mathbb{R}$, pointwise in t, uniformly with respect to δ.

Extracting a subsequence $\{w^\delta\}$, we have

$$M_2^\delta \rightharpoonup \tilde{M} \text{ as } \delta \downarrow 0, \tag{6.120}$$

weakly in the space of measures on $\mathbb{R} \times \mathbb{R}_+$; from (6.119), $\tilde{M}$ satisfies (6.111). A successive subsequence is extracted, applying lemma 6.8 with $\mu^\delta = u^\delta$ and $p = 3$, obtaining

$$u^\delta(\cdot, t) \to \underline{u}(\cdot, t) \text{ as } \delta \downarrow 0 \tag{6.121}$$

in $L_3(\mathbb{R})$, uniformly with respect to t in bounded intervals.

A final successive subsequence is extracted, again applying lemma 6.8 with $\mu^\delta = \nu^\delta$ and $p = 1$, obtaining

$$\nu^\delta(\cdot, t) \to \underline{\nu}(\cdot, t) \text{ as } \delta \downarrow 0$$

in $L_1(\mathbb{R})$, uniformly with resepct to t in bounded intervals. By appeal to lemma 6.8, in particular (6.107), the first space and time derivatives of the limit functions in (6.110) are of the claimed regularity.

It remains to prove (6.112) and (6.113), accomplished by evaluating the corresponding expressions for the approximations w^δ and passing to the limit $\delta \downarrow 0$, using lemma 6.8. From the specific, piecewise constant form of w^δ we have

$$u_t^\delta = -\sum_i s_i(u_{i,+} - u_{i,-})\delta(x - x_i), \tag{6.122}$$

$$(u^\delta)_x^2 = \sum_i (u_{i,+}^2 - u_{i,-}^2)\delta(x - x_i), \tag{6.123}$$

$$\tilde{v}_x^\delta = \sum_i (v_{i,+} - v_{i,-})\delta(x - x_i), \tag{6.124}$$

so for θ continuous and of compact support in $\mathbb{R} \times \mathbb{R}_+$

$$\begin{aligned} &\iint \theta\big(u_t^\delta + ((u^\delta)^2 - \tilde{v}^\delta)_x\big)dx\,dt \\ &= \int \sum_i \theta(x_i(t), t)\big[-s_i(u_{i,+} - u_{i,-}) + u_{i,+}^2 - u_{i,-}^2 - v_{i,+} + v_{i,-}\big]dt. \end{aligned} \tag{6.125}$$

In (6.125), the sum includes both singular shocks and jump discontinuities (6.85) satisfying (6.93). In both cases, the expression in brackets in (6.125) vanishes, using (6.29) for the singular shocks and (6.90), (6.94) for the jump discontinuities.

Thus the left side of (6.125) vanishes for all such θ. Replacing $\tilde{v}^\delta$ from (6.97), we have

$$u_t^\delta + \big(\frac{1}{2}(u^\delta)^2 + \nu^\delta\big)_x = 0 \tag{6.126}$$

weakly in the space of measures on $\mathbb{R}$. By appeal to lemma 6.8, we pass to the limit as $\delta \downarrow 0$, obtaining

$$\underline{u}_t + (\frac{1}{2}\underline{u}^2 + \underline{\nu})_x = 0 \tag{6.127}$$

which is equivalent to (1.112), using (1.110).

To prove (1.113), we use

$$\begin{aligned} \frac{\partial \tilde{M}_2^\delta}{\partial t} &= \sum_i \frac{\partial}{\partial t}\big(M_{i,2}\,\delta(x - x_i)\big) \\ &= \sum_i \Big(\frac{dM_{i,2}}{dt}\delta(x - x_i) - M_{i,2}\,s_i\,\delta_x(x - x_i)\Big) \\ &= \sum_i \big(e_2\delta(x - x_i) - M_{i,2}\,s_i\,\delta_x(x - x_i)\big), \end{aligned} \tag{6.128}$$

in which the sum is over the singular shocks. In (6.128), e_2 is the second component of the Rankine-Hugoniot deficit, given in (6.91), and $M_{i,2}$ is the second component of the singular mass M_i. We have also used (6.104) in obtaining (6.128).

From the form of $\tilde{v}^\delta$ and u^δ

$$\tilde{v}_t^\delta = -\sum_i s_i(v_{i,+} - v_{i,-})\delta(x - x_i) \tag{6.129}$$

and

$$\big((u^\delta)^3\big)_x = \sum_i (u_{i,+}^3 - u_{i,-}^3)\delta(x - x_i) \tag{6.130}$$

with the sums in (1.129) and (1.130) including both singular shocks and jump discontinuities.

Thus using (6.128-6.130), for θ as in (6.113)

$$\iint \theta((\tilde{v}^\delta + \tilde{M}_2^\delta)_t + \frac{1}{3}((u^\delta)^3)_x)dx\,dt$$
$$= \int \sum_i \theta(x_i(t),t)[-s_i(v_{i,+} - v_{i,-}) + \frac{1}{3}u_{i,+}^3 - u_{i,-}^3]dt$$
$$+ \int \sum_j \theta(x_j(t),t)[e_{2,j} - s_j(v_{j,+} - v_{j,-}) + \frac{1}{3}u_{j,+}^3 - \frac{1}{3}u_{j,-}^3]dt$$
$$(6.131) \qquad + \int \sum_j \theta_x(x_j(t),t)s_j M_{j,2}\,dt$$

in which the $\sum_i$ includes only the jump discontinuities and the $\sum_j$ includes only the singular shocks.

Using (6.91), (6.94) and (6.117), the expression in brackets in the first right-hand term in (6.131) is $(u_{i,+} - u_{i,-})^3/12$. Using (6.93), (6.95) and (6.114), the first right-hand term in (6.131) is $O(\delta^2)$.

The second right hand term in (6.131) vanishes, using (6.91) and (6.117).

For θ as in (6.113), the amount of singular mass outside of the region in which θ_x vanishes is $o(1)$ as $\delta \downarrow 0$.

Thus the left side of (6.131) is $o(1)$ as $\delta \downarrow 0$. Passage to the limit $\delta \downarrow 0$ is again justified by appeal to lemma 6.8, thus obtaining (6.113). □

The uniqueness of the solution obtained in this manner is unclear for several reasons. The solution is obtained as the limit of a convergent subsequence of approximations, not the entire sequence. The approximation of the initial data by a set of waves within the adopted set $\mathcal{A}_\delta$ is not unique. Finally, the individual approximations might not continue uniquely, as discussed in the proof of lemma 6.6.

A simple condition on the initial data assures uniqueness of the solution obtained by such a front-tracking algorithm, the above ambiguities notwithstanding. The proof shows, however, that the class of solutions obtainable in this manner is quite restricted.

THEOREM 6.10. *For the system (2.24, 2.25), assume initial data of bounded variation, with $\nu(\cdot,0)$ piecewise constant, satisfying (6.99), and such that there are no accumulation points of the set $\{\underline{x}_i(0)\}$ of points at which $\nu(\cdot,0)$ is discontinuous. The the solution obtained by application of a front-tracking algorithm, using a wave set $\mathcal{A}_\delta$ containing the singular shocks satisfying (6.28), (6.29) and the jump discontinuities satisfying (6.93), is unique.*

PROOF. By examining each of the separate cases in the proof of theorem 6.4, we observe that given any initial value problem (5.1) or (5.2) for this system with $\nu_r = \nu_\ell$, then any intermediate state w_m appearing in the solution will satisfy $\nu_m = \nu_r = \nu_\ell$. A front-tracking algorithm with this wave set $\mathcal{A}_\delta$ thus preserves constant ν.

Given such initial data, a singular shock with limiting values $\nu_+ \neq \nu_-$ will form immediately at each of the points $\underline{x}_i(0)$, in each of the front-tracking approximations w^δ. Below we refer to these singular shocks with $\nu_+ \neq \nu_-$ as essential.

By appeal to theorem 6.4, these essential singular shocks cannot fission as t increases. Of course they may collide, resulting in a single essential singular shock, or disappearing as such if ν_+ becomes equal to ν_- as the result of the collision. As $|u^\delta|$ is uniformly bounded, the speeds of the singular shocks are, from (6.28), so the essential singular shocks, locally finite in density at $t = 0$, remain so for all later t.

Thus for each $\delta > 0$, the essential singular shocks in w^δ divide $\mathbb{R} \times \mathbb{R}_+$ into disjoint regions, each of which borders the line $t = 0$. Within each of these regions, ν^δ is constant, determined by its value at $t = 0$.

Again by appeal to theorem 6.4, the front-tracking approximations satisfy

$$u^\delta(x+0,t) - u^\delta(x-0,t) \leq \delta \tag{6.132}$$

using (6.93).

The limits(s) of such sequences of approximations inherit similar structure, and satisfy the conclusions of theorem 6.9. The function(s) $\underline{\nu}$ are piecewise constant, determined by the initial data, in each of the disjoint regions separated by the essential singular shocks appearing in the solution. Within each such region, from (6.12) and (6.110), $\underline{u}$ satisfies Burgers equation

$$\underline{u}_t + (\frac{1}{2}\underline{u}^2)_x = 0, \tag{6.133}$$

weakly in the space of measures on $\mathbb{R} \times \mathbb{R}_+$. Since $\underline{u}_x(\cdot, t)$ is a bounded measure on $\mathbb{R}$, uniformly with respect to t, from (6.132) $\underline{u}_x$ satisfies a familiar entropy condition

$$\underline{u}_x < \infty \tag{6.134}$$

in the space of measures on $\mathbb{R}$, for each $t > 0$. Thus $\underline{u}$ is an entropy solution of (6.133), and is uniquely determined in any open subset R of $\mathbb{R} \times \mathbb{R}_+$ by the initial data and the value of $\underline{u}$ on any characteristics entering R as t increases.

Let R denote such a region, bounded by the initial line and essential singular shock carves for $\underline{w}$. The characteristics for (6.133) satisfy

$$\frac{dx}{dt} = \underline{u}(x,t) \tag{6.135}$$

so the entropy condition (6.28) satisfied by the essential singular shocks associated with a solution $\underline{w}$, does not preclude characteristic curves separating from the shock curves as t increases. Indeed this can happen, but only at points at which the shock speed s is characteristic, equal to the limiting value of $\underline{u}$ on one side. Thus given the shock speed s as a function of t, the values of $\underline{u}$ on any characteristics entering R are known. It follows that $\underline{u}$ is uniquely determined throughout $\mathbb{R} \times \mathbb{R}_+$ by the initial data and the essential singular shock curves $\{\underline{x}_i(t),\ t > 0,\ i = 1, \cdots\}$.

The function $\underline{\nu}$ is determined within each such region R by the initial data. Thus $\underline{\nu}$ is uniquely determined throughout $\mathbb{R} \times \mathbb{R}_+$ by the essential singular shock curves.

Given $\underline{u}$, using (6.13) and the initial data, $\underline{v}$ is uniquely determined throughout $\mathbb{R}\times\mathbb{R}_+$. Thus to prove uniqueness of $\underline{w}$ it suffices to prove that the essential singular shock curves continue uniquely at each point, i.e. that s is uniquely determined.

At $t = 0$, the value of s at each point $\underline{x}_i(0)$ is uniquely obtained from the solution of the corresponding Riemann problem, by appeal to theorem 6.4.

At points $\underline{x}_i(t)$, $t > 0$, it suffices to establish that s is uniquely determined from (6.28-6.30), given the limiting values $\underline{\nu}_\pm$ (which one determined by the initial data) and the values of $\underline{u}_\pm$ on the characteristics incoming to the shock.

For an essential singular shock $|\underline{\nu}_+ - \underline{\nu}_-| > 0$, so $\underline{u}_+ < \underline{u}_-$ from (6.30). If (6.28) holds with strict inequality, $\underline{u}_+ < s < \underline{u}_-$, then both characteristics are incoming, so both $\underline{u}_\pm$ are known and s is determined explicitly from (6.29), using (2.26),

$$s = \frac{1}{2}\left(\underline{u}_- + \underline{u}_+\right) + \frac{\underline{\nu}_- - \underline{\nu}_+}{\underline{u}_- - \underline{u}_+}. \tag{6.136}$$

Assume for definiteness that at some value t_0,

$$s(t_0) = \underline{u}_+(t_0) < \underline{u}_-(t_0). \tag{6.137}$$

Then for t in a neighborhood of t_0, the characteristic from the left is incoming, so $\underline{u}_-(t)$ is known; indeed it is lower semicontinuous from (6.134).

In this situation (6.30) holds with equality,

$$\underline{\nu}_+ - \underline{\nu}_- = \frac{1}{2}\left(\underline{u}_- - \underline{u}_+\right)^2, \tag{6.138}$$

so if $\underline{u}_-$ is decreasing as a function of t, in a neighborhood of t_0, then $\underline{u}_+(t)$ must depend on $\underline{u}_-(t)$ to maintain (6.30).

Such a situation arises as the limit of the interaction of a singular shock of characteristic speed, satisfying (6.137), with jump discontinuities (6.85), (6.93) approaching from the left. Theorem 6.4 uniquely determines the result of such interactions; s, u_- and u_+ for the singular shock decrease, leaving $s = u_+$ and $\nu_\pm$ and $u_- - u_+$ unaffected, so that (6.30) remains satisfied. Thus in the limit, $\underline{u}_+$ and s are indeed uniquely determined as functions of t by $\underline{u}_-(t)$, so that $s = \underline{u}_+$ and (6.138) holds at each t in a neighborhood of t_0.

As soon as $\underline{u}_-$ increases with t, s increases while $\underline{u}_+$ remains fixed, so both characteristics are incoming and s is obtained from (6.136). □

Bibliography

[AH] R. K. Agarwal, D. W. Halt, *A modified CUSP scheme in wave/particle split form for unstructured grid Euler flows*, in *Frontiers of Computational Fluid Dynamies*, D. A. Caughey and M. M. Hafes, eds., Wiley, 1994.

[B] F. Bouchut, *On zero-pressure gas dynamics*, in *Advances in kenetic theory and computing, Series on Advances in Mathematics for Applied Sciences*, vol. 122 World Scientific, pp. 171–190.

[BCP] A. Bressan, G. Crasta and P. Piccoli, *Well-Posedness of the Cauchy problem for $n \times n$ Systems of Conservation Laws*, A.M.S. Memoirs #694, American Mathematical Society, Providence, R.I. (2000).

[BG] Y. Brenier and E. Grenier, *Sticky particles and scalar conservation laws*, SIAM J. Num. Anal. **35** (1998), pp. 2317–2328.

[BJ] P. Baiti and H. K. Jenssen, *On the front-tracking algorithm*, J. Math. Anal. Appl. **217** (1998), pp. 395–404.

[BK] G. W. Bluman and S. Kumei, *Symmetries and differential Equations*, Applied Mathematical Sciences vol. 81, Springer-Verlag, New York, 1989.

[Br] A. Bressan, *Global solutions of systems of conservation laws by wave-front tracking*, J. Math. Anal. Appl. **170** (1992), pp. 414–432.

[CCL] J. J. Cauret, J. F. Colombeau, and A. Y. Le Roux, *Discontinuous generalized solutions of nonlinear nonconservative hyperbolic equations*, J. Math. Anal. Appl. **139** (1989), pp. 552–573.

[CL1] J. F. Colombeau and A. Y. Le Roux, *Numerical techniques in elastoplasticity*, in: *Nonlinear Hyperbolic Problems*, C. Carasso, P. Raviart and D. Serre, eds., Lecture Notes in Mathematics **1270** (1987) Springer-Verlag, Berlin, pp. 103–114.

[CL2] J. F. Colmbeau and A. Y. Le Roux, *Multiplications of distributions in elasticity and hydrodynamics*, J. Math. Phys. **29** (1988) pp. 315–319.

[CLZ] S. Cheng, J. Li and J. Zhang, *Explicit construction of measure solutions of the Cauchy problem for the transportation equations*, Science in China, Series A **40** (1997), pp. 1287–1299.

[D] C. M. Dafermos, *Hyperbolic Conservation Laws in Continuum Physics*, Springer-Verlag, Berlin-Heidelberg, 2000.

[DD] C. M. Dafermos and R. J. DiPerna, *The Riemann problem for certain classes of hyperbolic systems of conservation laws*, J. Diff. Eq. **20** (1976), pp. 90–114.

[ERS] Weinan E, Y. C. Rykov and Ya. G. Sinai, *Generalized variational principles, global weak solutions and behavior with random initial data for systems of conservation laws arising in adhesion particle dynamics*, Comm. Phys. Math. **177** (1996), pp. 349–380.

[F] L. R. Foy, *Steady state solutions of hyperbolic systems of conservation laws with viscous terms*, Comm. Pure Appl. Math. **17** (1964), pp. 177–188.

[FL] K. O. Friedrichs and P. D. Lax, *Systems of conservation laws with a convex extension*, Proc. Nat. Acad. Sci. USA **68** (1971) pp. 1686–1688.

[G] J. Glimm, *Solutions in the large for nonlinear hyperbolic systems of equations*, Comm. Pure Appl. Math. **18** (1965) pp. 95–105.

[Ge] I. Gelfand, *Some problems in the theory of quasilinear equations*, Usp. Mat. Nauk **14** (1959) pp. 87–158; Amer. Math. Soc. Transl. series 2 (1963) pp. 295–381.

[Go] S. K. Godunov, *An interesting class of quasilinear systems*, Dokl. Akad. Nauk SSSR **139** (1961) pp. 521–523.

[Ke] B. L. Keyfitz, *Conservation laws, delta shocks and singular shocks*, in: *Nonlinear Theory of Generalized Functions*, M. Grosser, G. Hormann, M. Kunzinger and M. Oberguggenberger, eds., Chapman and Hall/CRC Press, Boca Raton (1999) pp. 99–111.

[KK1] B. L. Keyfitz and H. C. Kranzer, *A system of hyperbolic conservation laws arising in elasticity theory*, Arch. Rat. Mech. Anal. **72** (1980), pp. 219–241.

[KK2] B. L. Keyfitz and H. C. Kranzer, *Spaces of weighted measures for conservation laws with singular solutions*, J. Diff. Eq. **118** (1995) pp. 420–451.

[KK3] B. L. Keyfitz and H. C. Kranzer, *A viscous approximation to a system of conservation laws with no classical Riemann solution*, in *Nonlinear Hyperbolic Problems*, C. Carasso, ed., Lecture Notes in Mathematics vol. 1402, pp. 185–197, Springer-Verlag, Berlin/New York, 1989.

[KK4] B. L. Keyfitz and H. C. Kranzer, *A system of conservation laws with no classical Riemann solution*, Univ. Houston Math. Dept. preprint UH/MD86, 1990.

[KK5] H. C. Kranzer and B. L. Keyfitz, *A strictly hyperbolic system of conservation laws admitting singular shocks*, in *Nonlinear Evolution Equations that Change Type*, B. Keyfitz and M. Shearer, eds., IMA vol. Math. Appl. Vol 27, pp. 107–125, Springer-Verlag, Berlin/New York (1990).

[KK6] B. L. Keyfitz and H. C. Kranzer, *Existence and uniqueness of entropy solutions to the Riemann problem for hyperbolic systems of two nonlinear conservation laws*, J. Diff. Eq. **27** (1978) pp. 444–476.

[KK7] B. L. Keyfitz and H. C. Kranzer, *The Riemann problem for a class of hyperbolic conservation laws exhibiting a parabolic degeneracy*, J. Diff. Eq. **47** (1983), pp. 35–65.

[Kl] D. Klein, *Existence of Convex Entropy Densities for some Strictly Hyperbolic Pairs of Conservation Laws*, Ph.D. thesis, Mathematics Department, The Hebrew University of Jerusalem (1994).

[KLS] B. L. Keyfitz, M. Lewicka and M. Sever, *The Cauchy problem for a model equation with singular shocks*, preprint.

[Ko] D. J. Korchinski, *Solution of a Riemann Problem for a 2×2 System of Conservation Laws possessing no Classical Weak Solution*, Ph.D. thesis, Adelphi Univ. Garden City, NY, 1977.

[KPS] L. Kofman, D. Pogosyan and S. Shandarin, *Structure of the universe in the two-dimensional model of adhesion*, Mon. Nat. R. Astr. Soc. **242** (1990), pp. 200–208.

[L1] P. D. Lax, *Hyperbolic systems of conservation laws, II*, Comm. Pure Appl. Math. **10** (1957), pp. 537–566.

[L2] P. D. Lax, *Shock waves and entropy*, Proc. Symp. Univ. of Wisconsin, E. H. Zarontonello, ed., Academic Press, New York, 1971, pp. 603–634.

[L3] P. D. Lax, *Hyperbolic Systems of Conservation Laws and the Mathematical Theory of Shock Waves*, SIAM Regional Conference Series in Applied Mathematics, 1973.

[LC] Y. Li and Y. Cao, *Large particle difference method with second order occuracy in gas dynamics*, Scientific Sinica (A) **28** (1985), ppo. 1024–1035.

[Lf] P. Le Floch, *An existence and uniqueness result for two nonstrictly hyperbolic systems*, in *Nonlinear evolution equations that change type*, B. L. Keyfitz and M. Shearer, eds., IMA Series #27, Springer-Verlag, 1990, pp. 126–138.

[Li] J. Li, *Note on the compressible Euler equations with zero temperature*, Preprint of Department of Mathematics, Otto-von-Guericke Universitat Magdeburg, 1999.

[Liu1] T. P. Liu, *The Riemann problem for general systems of conservation laws*, J. Diff. Eq. **18** (1975) pp. 218–234.

[Liu2] T. P. Liu, *The Riemann problem for 2×2 conservation laws*, Trans. Amer. Math. Soc. **199** (1974), pp. 89–112.

[Liu3] T. P. Liu, *Existence and uniqueness theorems for Riemann problems*, Trans. Amer. Math. Soc. **212** (1975), pp. 375–382.

[Liu4] T. P. Liu, *The entropy condition and admissibility of shocks*, J. Math. Anal. Appl. **53** (1976), pp. 78–88.

[LT] R. Levesgue and B. Temple, *Stability of Godunov's method for a class of 2×2 systems of conservation laws*, Trans. AMS **288** (1985), pp. 115–123.

[LW] J. Li and G. Warnecke, *On the uniqueness of entropy solutions to zero pressure gas dynamics*, preprint.

[LZ] J. Li and T. Zhang, *Generalized Rankine-Hugoniot relations of delta-shocks in solution of transportation equations*, in *Proc. Int. Conf. PDE*, G. Q. Chen, ed., (1997).

[MMZ] A. Majda, G. Majda and Y. Zheng, *Concentrations in the one-dimensional Vlasov-Poisson equations I: temporal development and non-unique weak solutions to the single component case*, Physica D **74** (1994), pp. 268–300.

[Mo1] M. Mock, *Systems of conservation laws of mixed type*, J. Diff. Eq. **37** (1980), pp. 70–88.

[Mo2] M. Mock, *A topological degree for orbits connecting critical points of automomous systems*, J. Diff. Eq. **38** (1980), pp. 176–191.

[N] T. Nishida, *Nonlinear hyperbolic equations and related topics in fluid dynamics*, Publications Matématiques D'Orsay, Septembre 1978.

[Ob] M. Oberguggenberger, *Case study of a nonlinear, nonconservative, nonstrictly hyperbolic system*, Nonlinear Analysis, Theory, Methods and Applications, **19** (1992), pp. 53–79.

[Ol] O. Oleinik, *On the uniqueness of the generalized solution of the Cauchy problem for a nonlinear system of equations occurring in mechanics*, Usp. Mat. Nauk **12** (1957), pp. 169–176.

[R] P. H. Rabinowitz, *Minimax Methods in Critical Point Theory with Applications to Differential Equations*, AMS Regional Conference Series in Mathematics, number 65 (1986).

[Ri] Nils Henrik Risebro, *A front-tracking alternative to the random choice method*, Proc. Amer. Math. Soc. **117** (1993), pp. 1125–1139.

[Ro] P. Rosenau, *Evolution and breaking of ion-accoustic waves*, Phys. Fluids **37** (1988), pp. 1317–1319.

[S1] M. Sever, *Existence in the large for Riemann problems for systems of conservation laws*, Trans. Amer. Math. Soc. **292** (1985), pp. 375–381.

[S2] M. Sever, *Exchange of conserved quantities, shock loci and Riemann problems*, Math. Methods in the Appl. Sci. **24** (2001), pp. 969–992.

[S3] M. Sever, *Exchange of conserved quantities in nonhyperbolic systems- an example*, Proceedings of the Amer. Math. Soc. **129** (2001), pp. 3671–3681.

[S4] M. Sever, *A class of nonlinear, nonhyperbolic systems of conservation laws with well-posed initial-value problems*, J. Diff. Eq. **180** (2002), pp. 238–271.

[S5] M. Sever, *Viscous structure of singular shocks*, Nonlinearity **15** (2002), pp. 705–725.

[S6] M. Sever, *Hyperbolic systems of conservation laws with some special invariance properties*, Israel J. Math. **75** (1991), pp. 81–104.

[S7] M. Sever, *Hyperbolic systems of conservation laws with a strict Riemann invariant*, J. Diff. Eq. **122** (1995), pp. 239–266.

[S8] M. Sever, *A variational formulation of symmetric systems of conservation laws*, Houston Journal of Mathematics **3** (2000), pp. 561–573.

[Se1] D. Serre, *Solution à variations bornées pour certains systèms hyperboliques de lois de conservation*, J. Diff. Eq. **68** (1987), pp. 137–168.

[Se2] D. Serre, *Intégrabilité d' une classe de systèms du lois de conservation*, Forum Mathematica **4** (1992), pp. 607–623.

[Se3] D. Serre, *Systèms hyperboliques riches de lois de conservation*, in: *Nonlinear PDEs and their applications*, College de France Seminars v.11, Pitman Research Notes #299, Longman (1994), pp. 248–281.

[SS] R. Sanders and M. Sever, Computations with singular shocks, SIAM J. Appl. Dynamical Systems. (to appear).

[SSS] D. G. Schaeffer, S. Schecter and M. Shearer, *Nonstrictly hyperbolic conservation laws with a parabolic line*, J. Diff. Eq. **103** (1993), pp. 94–126.

[SV] R. Saxton and V. Vinod, *Singularity formation in systems of nonstrictly hyperbolic equations*, Electr. J. Diff. Eq. **9** (1995), pp. 1–15.

[SZ] S. F. Shandarin and Y. B. Zeldovich, *The large-scale structure of the universe: Turbulence, intermittency, structures in a self-gravitating medium*, Rev. Modern Phys. **61** (1989), pp. 185–220.

[T1] B. Temple, *Systems of conservation laws with invariant submanifolds*, Trans. AMS **280** (1983), pp. 781–795.

[T2] B. Temple, *Solutions in the large for some nonlinear hyperbolic conservation laws of gas dynamics*, J. Diff. Eq. **41** (1981), pp. 96–161.

[TZ] D. Tan and T. Zhang, *Two-dimensional Riemann problem for a hyperbolic system of nonlinear conservation laws (I): four-J cases*, J. Diff. Eq. **111** (1994), pp. 203–254.

[TZZ] D. Tan, T. Zhang and Y. Zheng, *Delta-shock waves as limits of vanishing viscosity for hyprbolic system of conservation laws*, J. Diff. Eq. **112** (1994), pp. 1–32.

[W1] D. Wagner, *Equivalence of the Euler and Lagrangian equations of gas dynamics for weak solutions*, J. Diff. Eq. **68** (1987), pp. 118–136.

[W2] D. Wagner, *Conservation laws, coordinate transformations and differential forms*, in *proceedings of the Fifth International Conference on Hyperbolic Problems: Theory, Numerics, and Applications*, J. Glimm, M. J. Graham, J. W. Grove, adn B. J. Plohr, eds., World Scientific (1994), pp. 471–477.

[YL] H. Yang and J. Li, *Delta-shocks as limits of vanishing viscosity for multidimensional zero-pressure gas dynamics*, Q. Appl. Math. (to appear).

[Z] Y. Zheng, *Systems of conservation laws with incomplete sets of eigenvalues everywhere* in *Advances in Nonlinear Partial Differential Equations and Related Areas*, Gui-Qiang Chen et. al, eds., World Scientific Singapore, 1998.

Editorial Information

To be published in the *Memoirs*, a paper must be correct, new, nontrivial, and significant. Further, it must be well written and of interest to a substantial number of mathematicians. Piecemeal results, such as an inconclusive step toward an unproved major theorem or a minor variation on a known result, are in general not acceptable for publication.

Papers appearing in *Memoirs* are generally at least 80 and not more than 200 published pages in length. Papers less than 80 or more than 200 published pages require the approval of the Managing Editor of the Transactions/Memoirs Editorial Board.

As of July 31, 2007, the backlog for this journal was approximately 15 volumes. This estimate is the result of dividing the number of manuscripts for this journal in the Providence office that have not yet gone to the printer on the above date by the average number of monographs per volume over the previous twelve months, reduced by the number of volumes published in four months (the time necessary for preparing a volume for the printer). (There are 6 volumes per year, each usually containing at least 4 numbers.)

A Consent to Publish and Copyright Agreement is required before a paper will be published in the *Memoirs*. After a paper is accepted for publication, the Providence office will send a Consent to Publish and Copyright Agreement to all authors of the paper. By submitting a paper to the *Memoirs*, authors certify that the results have not been submitted to nor are they under consideration for publication by another journal, conference proceedings, or similar publication.

Information for Authors

Memoirs are printed from camera copy fully prepared by the author. This means that the finished book will look exactly like the copy submitted.

Initial submission. The AMS uses Centralized Manuscript Processing for initial submissions. Authors should submit a PDF file using the Initial Manuscript Submission form found at `www.ams.org/cgi-bin/peertrack/submission.pl`, or send one copy of the manuscript to the following address: Centralized Manuscript Processing, MEMOIRS OF THE AMS, 201 Charles Street, Providence, RI 02904-2294 USA. If a paper copy is being forwarded to the AMS, indicate that it is for it Memoirs and include the name of the corresponding author, contact information such as email address or mailing address, and the name of an appropriate Editor to review the paper (see the list of Editors below).

The paper must contain a *descriptive title* and an *abstract* that summarizes the article in language suitable for workers in the general field (algebra, analysis, etc.). The *descriptive title* should be short, but informative; useless or vague phrases such as "some remarks about" or "concerning" should be avoided. The *abstract* should be at least one complete sentence, and at most 300 words. Included with the footnotes to the paper should be the 2000 *Mathematics Subject Classification* representing the primary and secondary subjects of the article. The classifications are accessible from `www.ams.org/msc/`. The list of classifications is also available in print starting with the 1999 annual index of *Mathematical Reviews*. The Mathematics Subject Classification footnote may be followed by a list of *key words and phrases* describing the subject matter of the article and taken from it. Journal abbreviations used in bibliographies are listed in the latest *Mathematical Reviews* annual index. The series abbreviations are also accessible from `www.ams.org/publications/`. To help in preparing and verifying references, the AMS offers MR Lookup, a Reference Tool for Linking, at `www.ams.org/mrlookup/`.

Electronically prepared manuscripts. The AMS encourages electronically prepared manuscripts, with a strong preference for $\mathcal{A}_{\mathcal{M}}\mathcal{S}$-LaTeX. To this end, the Society has prepared $\mathcal{A}_{\mathcal{M}}\mathcal{S}$-LaTeX author packages for each AMS publication. Author packages include instructions for preparing electronic manuscripts, samples, and a style file that generates the particular design specifications of that publication series. Though $\mathcal{A}\mathcal{M}\mathcal{S}$-LaTeX is the highly preferred format of TeX, author packages are also available in $\mathcal{A}\mathcal{M}\mathcal{S}$-TeX.

Authors may retrieve an author package from the AMS website starting from `www.ams.org/tex/` or via FTP to `ftp.ams.org` (login as `anonymous`, enter username as password, and type `cd pub/author-info`). The *AMS Author Handbook* and the *Instruction Manual* are available in PDF format following the author packages link from `www.ams.org/tex/`. The author package can also be obtained free of charge by sending email to `tech-support@ams.org` (Internet) or from the Publication Division, American Mathematical Society, 201 Charles St., Providence, RI 02904-2294, USA. When requesting an author package, please specify $\mathcal{A}\mathcal{M}\mathcal{S}$-LaTeX or $\mathcal{A}\mathcal{M}\mathcal{S}$-TeX and the publication in which your paper will appear. Please be sure to include your complete mailing address.

After acceptance. The final version of the electronic file should be sent to the Providence office (this includes any TeX source file, any graphics files, and the DVI or PostScript file) immediately after the paper has been accepted for publication.

Before sending the source file, be sure you have proofread your paper carefully. The files you send must be the EXACT files used to generate the proof copy that was accepted for publication. For all publications, authors are required to send a printed copy of their paper, which exactly matches the copy approved for publication, along with any graphics that will appear in the paper.

Accepted electronically prepared files can be submitted via the web at `www.ams.org/submit-book-journal/`, sent via FTP, or sent on CD-Rom or diskette to the Electronic Prepress Department, American Mathematical Society, 201 Charles Street, Providence, RI 02904-2294 USA. TeX source files, DVI files, and PostScript files can be transferred over the Internet by FTP to the Internet node `ftp.ams.org` (130.44.1.100). When sending a manuscript electronically via CD-Rom or diskette, please be sure to include a message identifying the paper as a Memoir.

Electronically prepared manuscripts can also be sent via email to `pub-submit@ams.org` (Internet). In order to send files via email, they must be encoded properly. (DVI files are binary and PostScript files tend to be very large.)

Electronic graphics. Comprehensive instructions on preparing graphics are available at `www.ams.org/jourhtml/`. A few of the major requirements are given here.

Submit files for graphics as EPS (Encapsulated PostScript) files. This includes graphics originated via a graphics application as well as scanned photographs or other computer-generated images. If this is not possible, TIFF files are acceptable as long as they can be opened in Adobe Photoshop or Illustrator. No matter what method was used to produce the graphic, it is necessary to provide a paper copy to the AMS.

Authors using graphics packages for the creation of electronic art should also avoid the use of any lines thinner than 0.5 points in width. Many graphics packages allow the user to specify a "hairline" for a very thin line. Hairlines often look acceptable when proofed on a typical laser printer. However, when produced on a high-resolution laser imagesetter, hairlines become nearly invisible and will be lost entirely in the final printing process.

Screens should be set to values between 15% and 85%. Screens which fall outside of this range are too light or too dark to print correctly. Variations of screens within a graphic should be no less than 10%.

Inquiries. Any inquiries concerning a paper that has been accepted for publication should be sent to `memo-query@ams.org` or directly to the Electronic Prepress Department, American Mathematical Society, 201 Charles St., Providence, RI 02904-2294 USA.

Editors

This journal is designed particularly for long research papers, normally at least 80 pages in length, and groups of cognate papers in pure and applied mathematics. Papers intended for publication in the *Memoirs* should be addressed to one of the following editors. The AMS uses Centralized Manuscript Processing for initial submissions to AMS journals. Authors should follow instructions listed on the Initial Submission page found at www.ams.org/memo/memosubmit.html.

Algebra to ALEXANDER KLESHCHEV, Department of Mathematics, University of Oregon, Eugene, OR 97403-1222; email: ams@noether.uoregon.edu

Algebraic geometry and its application to MINA TEICHER, Emmy Noether Research Institute for Mathematics, Bar-Ilan University, Ramat-Gan 52900, Israel; email: teicher@macs.biu.ac.il

Algebraic geometry to DAN ABRAMOVICH, Department of Mathematics, Brown University, Box 1917, Providence, RI 02912; email: amsedit@math.brown.edu

Algebraic number theory to V. KUMAR MURTY, Department of Mathematics, University of Toronto, 100 St. George Street, Toronto, ON M5S 1A1, Canada; email: murty@math.toronto.edu

Algebraic topology to ALEJANDRO ADEM, Department of Mathematics, University of British Columbia, Room 121, 1984 Mathematics Road, Vancouver, British Columbia, Canada V6T 1Z2; email: adem@math.ubc.ca

Combinatorics to JOHN R. STEMBRIDGE, Department of Mathematics, University of Michigan, Ann Arbor, Michigan 48109-1109; email: FRS@umich.edu

Complex analysis and harmonic analysis to ALEXANDER NAGEL, Department of Mathematics, University of Wisconsin, 480 Lincoln Drive, Madison, WI 53706-1313; email: nagel@math.wisc.edu

Differential geometry and global analysis to LISA C. JEFFREY, Department of Mathematics, University of Toronto, 100 St. George St., Toronto, ON Canada M5S 3G3; email: jeffrey@math.toronto.edu

Dynamical systems and ergodic theory to AMIE WILKINSON, Department of Mathematics, Northwestern University, 2033 Sheridan Road, Evanston, IL 60208-2730; email: transactions@math.northwestern.edu

Functional analysis and operator algebras to DIMITRI SHLYAKHTENKO, Department of Mathematics, University of California, Los Angeles, CA 90095; email: shlyakht@math.ucla.edu

Geometric analysis to WILLIAM P. MINICOZZI II, Department of Mathematics, Johns Hopkins University, 3400 N. Charles St., Baltimore, MD 21218; email: trans@math.jhu.edu

Geometric analysis to MLADEN BESTVINA, Department of Mathematics, University of Utah, 155 South 1400 East, JWB 233, Salt Lake City, Utah 84112-0090; email: bestvina@math.utah.edu

Harmonic analysis, representation theory, and Lie theory to ROBERT J. STANTON, Department of Mathematics, The Ohio State University, 231 West 18th Avenue, Columbus, OH 43210-1174; email: stanton@math.ohio-state.edu

Logic to STEFFEN LEMPP, Department of Mathematics, University of Wisconsin, 480 Lincoln Drive, Madison, Wisconsin 53706-1388; email: lempp@math.wisc.edu

Partial differential equations to GUSTAVO PONCE, Department of Mathematics, South Hall, Room 6607, University of California, Santa Barbara, CA 93106; email: ponce@math.ucsb.edu

Partial differential equations and dynamical systems to PETER POLACIK, School of Mathematics, University of Minnesota, Minneapolis, MN 55455; email: polacik@math.umn.edu

Probability and statistics to KRZYSZTOF BURDZY, Department of Mathematics, University of Washington, Box 354350, Seattle, Washington 98195-4350; email: burdzy@math.washington.edu

Real analysis and partial differential equations to DANIEL TATARU, Department of Mathematics, University of California, Berkeley, Berkeley, CA 94720; email: tataru@math.berkeley.edu

All other communications to the editors should be addressed to the Managing Editor, ROBERT GURALNICK, Department of Mathematics, University of Southern California, Los Angeles, CA 90089-1113; email: guralnic@math.usc.edu.

Titles in This Series

890 **Steven Dale Cutkosky,** Toroidalization of dominant morphisms of 3-folds, 2007

889 **Michael Sever,** Distribution solutions of nonlinear systems of conservation laws, 2007

888 **Roger Chalkley,** Basic global relative invariants for nonlinear differential equations, 2007

887 **Charlotte Wahl,** Noncommutative Maslov index and eta-forms, 2007

886 **Robert M. Guralnick and John Shareshian,** Symmetric and alternating groups as monodromy groups of Riemann surfaces I: Generic covers and covers with many branch points, 2007

885 **Jae Choon Cha,** The structure of the rational concordance group of knots, 2007

884 **Dan Haran, Moshe Jarden, and Florian Pop,** Projective group structures as absolute Galois structures with block approximation, 2007

883 **Apostolos Beligiannis and Idun Reiten,** Homological and homotopical aspects of torsion theories, 2007

882 **Lars Inge Hedberg and Yuri Netrusov,** An axiomatic approach to function spaces, spec tral synthesis and Luzin approximation, 2007

881 **Tao Mei,** Operator valued Hardy spaces, 2007

880 **Bruce C. Berndt, Geumlan Choi, Youn-Seo Choi, Heekyoung Hahn, Boon Pin Yeap, Ae Ja Yee, Hamza Yesilyurt, and Jinhee Yi,** Ramanujan's forty identities for Rogers-Ramanujan functions, 2007

879 **O. García-Prada, P. B. Gothen, and V. Muñoz,** Betti numbers of the moduli space of rank 3 parabolic Higgs bundles, 2007

878 **Alessandra Celletti and Luigi Chierchia,** KAM stability and celestial mechanics, 2007

877 **María J. Carro, José A. Raposo, and Javier Soria,** Recent developments in the theory of Lorentz spaces and weighted inequalities, 2007

876 **Gabriel Debs and Jean Saint Raymond,** Borel liftings of Borel sets: Some decidable and undecidable statements, 2007

875 **C. Krattenthaler and T. Rivoal,** Hypergéométrie et fonction zêta de Riemann, 2007

874 **Sonia Natale,** Semisolvability of semisimple Hopf algebras of low dimension, 2007

873 **A. J. Duncan,** Exponential genus problems in one-relator products of groups, 2007

872 **Anthony V. Geramita, Tadahito Harima, Juan C. Migliore, and Yong Su Shin,** The Hilbert function of a level algebra, 2007

871 **Pascal Auscher,** On necessary and sufficient conditions for L^p-estimates of Riesz transforms associated to elliptic operators on $\mathbb{R}^n$ and related estimates, 2007

870 **Takuro Mochizuki,** Asymptotic behaviour of tame harmonic bundles and an application to pure twistor D-modules, Part 2, 2007

869 **Takuro Mochizuki,** Asymptotic behaviour of tame harmonic bundles and an application to pure twistor D-modules, Part 1, 2007

868 **Gelu Popescu,** Entropy and multivariable interpolation, 2006

867 **Vilmos Totik,** Metric properties of harmonic measures, 2006

866 **William Craig,** Semigroups underlying first-order logic, 2006

865 **Nathanial P. Brown,** Invariant means and finite representation theory of $C*$-algebras, 2006

864 **John M. Lee,** Fredholm operators and Einstein metrics on conformally compact manifolds, 2006

863 **M. Lübke and A. Teleman,** The Universal Kobayashi-Hitchin correspondence on Hermitian manifolds, 2006

862 **Alberto Canonaco,** The Beilinson complex and canonical rings of irregular surfaces, 2006

861 **Leon A. Takhtajan and Lee-Peng Teo,** Weil-Petersson metric on the universal Teichmüller space, 2006